Online Journey Through™ Astronomy, Version 2.0: Stars and Galaxies

Michael Guidry and Margaret Riedinger

Student Companion

written by

Kevin Lee

University of Nebraska–Lincoln

THOMSON

™

BROOKS/COLE

Australia • Canada • Mexico • Singapore • Spain • United Kingdom • United States

THOMSON

BROOKS/COLE

Publisher: *David Harris*
Marketing Manager: *Erik Evans*
Acquisitions Editor: *Keith Dodson*
Advertising Project Manager: *Kelley McAllister*
Technology Project Manager: *Sam Subity*
Editorial Assistant: *Melissa Newt*
Cover Design: *Matt Perry*

Printed in the United States of America
1   2   3   4   5   6   7   08   07   06   05   04

Printer: Globus Printing

ISBN: 0-534-40003-5

For more information about our products, contact us at:
Thomson Learning Academic Resource Center
1-800-423-0563

For permission to use material from this text or product, submit a request online at
http://www.thomsonrights.com.
Any additional questions about permissions can be submitted by email to
thomsonrights@thomson.com.

Thomson Brooks/Cole
10 Davis Drive
Belmont, CA 94002-3098
USA

Asia
Thomson Learning
5 Shenton Way #01-01
UIC Building
Singapore 068808

Australia/New Zealand
Thomson Learning
102 Dodds Street
Southbank, Victoria 3006
Australia

Canada
Nelson
1120 Birchmount Road
Toronto, Ontario M1K 5G4
Canada

Europe/Middle East/South Africa
Thomson Learning
High Holborn House
50/51 Bedford Row
London WC1R 4LR
United Kingdom

Latin America
Thomson Learning
Seneca, 53
Colonia Polanco
11560 Mexico D.F.
Mexico

Spain/Portugal
Paraninfo
Calle/Magallanes, 25
28015 Madrid, Spain

Contents

Introduction

This companion contains numerous exercises to help you master the subject matter contained in *Online Journey Through™ Astronomy*. Although this web site presents a tremendous amount of information in a visually exciting and often interactive manner, only through using these astronomy concepts yourself to solve problems and draw conclusions will you gain a true understanding of them.

Each unit of the workbook is organized in the same manner. The first component is a list of *Chapter Objectives*. These objectives succinctly describe exactly what the authors would like students to learn. The major new vocabulary terms for the chapter are listed in a *Keywords* section. A *Progress Checklist* helps you keep track of which modules you have worked on, and you can check them off as you go. The fourth component is an *Introductory Narrative* describing the content of the chapter in a broad overview. The passage has blanks for you to fill in, most of which can be found in the Keywords section. The blanks in the passage could be filled in as you work through the chapter to reinforce vocabulary and concepts, or afterwards as a check of whether or not the major concepts have been retained. This section is designed to allow you an opportunity to use new vocabulary terms very early on in the learning process.

Exercises with various formats will follow depending on what is most suitable for the content of the chapter. Exercises involving graphing of data and drawing the appearance of objects from various geometrical perspectives will occur frequently. Performing simulations using Java™ Applets and analyzing the results will also be a common format. Some sections contain an additional exercise on a particularly important concept (Newton's Laws, for example). The first element of the exercise will typically already be completed to help students get started. Often more insight can be gained by viewing a related interactive component either during an exercise or immediately afterwards, and you will often be directed to do so. The interactive components will be referenced by the code IC followed by the label on the Master List of Animations (for example IC 18.2 would reference the Velocity Distribution Calculator).

Each unit concludes with a number of True/False questions. These questions will survey all of the ideas of the chapter and can be used as a final check of your mastery of the material.

Conceptual Maps are included at various intervals in this workbook. These sections tie together the material found in several chapters. The Conceptual Maps differ in appearance from other sections of the book in that they are large drawings that span two facing pages. There is typically an organizational structure such as a timeline or chart with blank labels and descriptions for you to fill in. The goal behind the Conceptual Maps is to show how the small pieces of knowledge mesh together to form the broad science of astronomy.

The appendix contains solutions to exercises. The answers to all of the Introductory Narratives, True/False Questions, and any exercises of a vocabulary or simulation nature are provided. Asterisks follow the names of any exercises for which solutions or feedback are supplied in the appendix. Solutions to graphical exercises are not provided due to space limitations.

The content, structure, and format of this Student Companion will evolve over time in an effort to better meet the needs of its users. Students and instructors with comments are encouraged to contact the author by e-mail at KLEE6@unl.edu.

Acknowledgments

The author would like to acknowledge the efforts of several people who aided in the development of the student companion. Michael Guidry and Margaret Riedinger participated in early discussions determining the structure and format of the Student Companion, and provided valuable guidance. Margaret Riedinger wrote the chapter objectives and keywords. She also read rough drafts of many of the exercises and suggested improvements.

Five students participated in a seminar class in introductory astronomy at the University of Nebraska–Lincoln and used early drafts of these exercises. Bhee J Arroyo, Scott Jefferson, Jeremy Kalina, Shauna Mullally, and Cody Pearson all provided feedback that ultimately improved this manuscript.

Unit 17
The Sun

Chapter Objectives

Our Sun is the only star we can study up close and in great detail; as such it will serve as the basis of comparison for all the other stars we will investigate. In this chapter we will list the basic properties of our Sun and its overall composition, and we will introduce the Standard Solar Model that attempts to explain these properties. The three directly observable layers of the Sun—the photosphere, chromosphere, and corona—will be described. The complex story of our Sun's changing magnetic field will be illustrated and the field's consequences, including the sunspot cycle and explosive solar flares, will be investigated.

Progress Checklist

1. Basic Properties
- ❏ Basic Solar Properties
- ❏ The Solar Composition
- ❏ The Interior of the Sun
- ❏ Helioseismology
- ❏ Standard Solar Model
- ❏ Solar Luminosity

2. Photosphere and Spectrum
- ❏ Imaging the Sun
- ❏ Photosphere
- ❏ Opacity
- ❏ The Solar Spectrum
- ❏ Granulation
- ❏ Chromosphere

3. Magnetic Field
- ❏ Sunspots
- ❏ Sunspot Cycle

- ❏ Active and Quiet Sun
- ❏ Zeeman Effect in Sunspots
- ❏ The Sun's Magnetic Field
- ❏ Origin of Sunspots

4. The Active Sun
- ❏ Active Regions
- ❏ Solar Prominences
- ❏ Solar Plages
- ❏ Solar Flares

5. Corona and Solar Wind
- ❏ The Solar Corona
- ❏ Corona and Solar Activity
- ❏ Coronal Holes and Solar Wind
- ❏ The Solar Wind
- ❏ Coronal Mass Ejections
- ❏ Influence of the Solar Wind

Keywords

photosphere
chromosphere
corona
convective zone
radiative zone
core
Frauenhofer lines
hydrogen
helium

helioseismology
velocity field
Standard Solar Model
hydrostatic equilibrium
solar luminosity
solar constant
SOHO
hydrogen-alpha
opaque

transparent
limb darkening
stellar opacity
optical depth
granulation
convection currents
supergranules
spicules
flash spectrum

sunspots	magnetic polarity	coronal mass ejections
umbra	magnetic cycle	quiescent prominence
penumbra	Maunder butterfly diagram	eruptive prominence
sunspot cycle	dynamo effect	plages
sunspot maximum	differential rotation	geomagnetic storms
sunspot minimum	magnetogram	coronagraph
active Sun	Babcock dynamo model	helmet streamers
quiet Sun	active regions	solar wind
Maunder Minimum	prominences	coronal holes
Zeeman effect	solar flares	

Exercise 17-1: Introductory Narrative

The Sun is the most studied star, and we believe it to be representative of all stars. The Sun is composed of a mixture of many gases, but it is mainly hydrogen and 1) _____. It is powered by nuclear reactions in its 2) _____. Energy moves to the surface through photons, a process known as radiative transport, and by the circulation of hot material, which is known as 3) _____. At all points in the Sun, the inward gravitational forces are balanced by outward gas and radiation pressure in a condition known as 4) _____. The 5) _____ of the Sun is the total amount of energy leaving the surface each second.

The visible surface of the Sun is called the 6) _____. This layer has sufficiently high density to produce enough light to be visible and sufficiently low density to allow this light to escape. Due to convection currents, this layer has a mottled appearance known as 7)_____. The region immediately above this layer is the 8) _____. Because it is very faint, it is commonly observed only during solar eclipses. The outer atmosphere of the Sun is called the 9) _____. It extends out into space many times the radius of the Sun and is much hotter than the surface of the Sun for reasons that are not well understood.

Many energetic and violent phenomena occur on the Sun and are indicative of periods of high solar activity. All of these phenomena appear to be related to the Sun's 10) _____ field. 11) _____ are regions on the surface of the Sun that appear dark because they are cooler than surrounding areas. A prominence is a stream of material shot out from the surface of the Sun that often falls backward forming a loop. More energetic eruptions from the surface are called solar 12) _____. These phenomena affect the Earth by increasing the flow of charged particles from the Sun known as the 13) _____.

Exercise 17-2: True/False Questions

T / F   1. The composition of the Sun is similar to that of the Universe as a whole.

T / F   2. Astronomers know the depth of the convective zone in the Sun from helio-seismology data.

T / F   3. The solar constant is the total amount of energy passing each second through the surface of a sphere surrounding the Sun with a radius of 1 AU.

T / F   4. The most important energy transport mechanism in the inner parts of the Sun is convection.

T / F   5. If one can see only a short distance through a fog, its opacity must be large.

T / F   6. The edge of the Sun appears brighter than the center because we are seeing light produced from hotter, deeper layers of the photosphere.

T / F   7. The chromosphere contains spikes of gas called spicules that may represent energy moving from the surface to the corona.

T / F   8. The Sun rotates more rapidly at the equator than it does near the poles.

T / F   9. The frequency of sunspots on the Sun remains fairly constant over time.

T / F 10. The presence of the Zeeman effect indicates that sunspots are magnetic phenomena.

T / F 11. The Babcock model explains the occurrence of sunspots as being due to especially strong convection.

T / F 12. Solar activity fluctuates with an 11-year period in the same way that sunspot counts fluctuate.

T / F 13. Coronal mass ejections increase the number and intensity of auroras on the Earth several days later.

Unit 18
Properties of the Stars

Chapter Objectives

In this chapter we learn that our Sun is a very average star that falls in the middle of the stellar range of possible luminosities, surface temperatures, and masses. We will study the laws of nuclear physics in order to understand how stars are able to release tremendous amounts of energy and yet remain so stable during their main sequence lifetimes. We will use stellar models to describe the energy transport that takes place in the hidden interiors of stars. We will show how astronomers determine the distances to stars and how they trace their motions in our sky. The H-R Diagram will be introduced as a graphical summary of some important characteristics of stars.

Progress Checklist

1. Energy Production
- ❏ Mass and Energy
- ❏ Curve of Binding Energy
- ❏ Nuclear Reactions
- ❏ Reaction Rates
- ❏ Temperature and Pressure
- ❏ The Energy Window

2. Stellar Burning Stages
- ❏ Hydrostatic Equilibrium
- ❏ Proton-Proton Chain
- ❏ The CNO Cycle
- ❏ PP-CNO Competition
- ❏ Triple-Alpha Process
- ❏ Advanced Burning Stages

3. Energy Transport
- ❏ Energy Transport & Equilibrium
- ❏ Radiative Transport
- ❏ Conduction
- ❏ Convection
- ❏ Neutrino Cooling
- ❏ Competition among Modes

4. Solar Neutrinos
- ❏ Neutrinos
- ❏ Detection of Neutrinos
- ❏ The Solar Neutrino Problem
- ❏ Resolution of the Problem

5. Stellar Distance and Motion
- ❏ The Parallax Method
- ❏ Units for Stellar Distances
- ❏ Distances to Nearby Stars
- ❏ Proper Motion
- ❏ Space Velocities
- ❏ Motion of the Sun

6. Stellar Magnitudes
- ❏ Magnitude Scale
- ❏ Apparent Magnitude
- ❏ Absolute Magnitude
- ❏ The Influence of Wavelength
- ❏ Astronomical Color Filters
- ❏ Color Indices

7. Harvard Spectral Sequence
- ❏ Spectral Sequence
- ❏ Origins and Remembrance
- ❏ Ionization
- ❏ Interpretation

8. HR Diagram
- ❏ HR Diagram
- ❏ Main Sequence
- ❏ Giants & Supergiants
- ❏ White Dwarfs
- ❏ Luminosity Classes
- ❏ Spectroscopic Parallax

Keywords

special theory of relativity
mass-energy conversion
thermonuclear
binding energy
nucleon
proton
neutron
nuclear fission
nuclear fusion
missing mass
Coulomb barrier
tunneling
kinetic theory of gases
Maxwellian distribution
Ideal Gas Law
Gamow window
hydrostatic equilibrium
Proton-Proton Chain
Carbon-Nitrogen-Oxygen cycle

Beta decay
positron
neutrino
deuterium
catalyst
Triple-Alpha process
radiative transport
conduction
convection
neutrino emission
degradation
random walk
degenerate matter
temperature gradient
solar neutrino problem
neutrino detectors
solar neutrino unit (SNU)
neutrino oscillation
parallax
parallax angle

parsec
light year
proper motion
space velocity
radial velocity
tangential velocity
solar apex
solar antapex
apparent magnitude
absolute magnitude
color index
Harvard Spectral Sequence
ionization
Hertzsprung-Russell Diagram
main sequence
red giant
white dwarf
supergiant
luminosity class
spectroscopic parallax

Exercise 18-1: Introductory Narrative (Stellar Energy Production)

Stars shine by converting mass into 1) _____. They do so through
2) _____ reactions in their cores. In these reactions two lighter nuclei combine to form a heavier nucleus. Because the heavier nuclei are more 3) _____,
nuclear energy is released in the process. Because the lighter nuclei are both
4) _____ charged, they repel each other. This electrical repulsion is known
as the 5) _____. The nuclei must be traveling very rapidly to get close
enough for the strong nuclear force to bind them together before the electrical repulsion
pushes them apart. Only in the core of stars is the 6) _____ high enough
that nuclei have the necessary velocities. In the most common reaction, two hydrogen nuclei
are converted into 7) _____. In low-mass stars, this occurs primarily by the
8) _____ chain and in high-mass stars mainly through the CNO Cycle.
The energy is transported from the core of the star to the surface mainly through radiation and
9) _____. Although very important in thermal transport on the Earth, conduction plays very little role in stars due to their low densities.

Exercise 18-2: Introductory Narrative (Stellar Parameters)

Astronomers are very interested in the intrinsic properties of stars such as distance, velocity, radius, mass, temperature, and composition. These parameters are interrelated in that knowing one of them tells you information about the others.

Distances are determined primarily through the method of 1) _____. This can be done only for stars within 100 pc using ground-based observations, but can be extended to about 2000 pc with observations from the 2) _____ satellite.

The space velocities of stars have two components that must be evaluated separately. The motion of the star on the celestial sphere is called 3) _____ and, if the distance to the star is known, can be used to calculate the tangential velocity. The radial velocity is calculated using the 4) _____. One can combine these two components trigonometrically to obtain the space velocity.

The perceived 5) _____ of a star is usually specified on a logarithmic system called apparent magnitude. This parameter does not take into account the 6) _____ of the star. The apparent magnitude of the star if it were moved to a distance of 10 pc is known as the 7) _____. This parameter describes the intrinsic brightness of the object.

The Harvard Spectral Sequence is used to describe the surface 8) _____ of a star. The sequence is denoted by the sequence of letters O, 9) _____, A, F, G, 10) _____, and M. It is determined by looking at the absorption lines in the spectrum of a star. Similar information is conveyed by the 11) _____ of a star, which is the difference between the magnitudes through two different filters.

A diagram of absolute magnitude or luminosity versus spectral type or color index is known as a(n) 12) _____ Diagram. The location of a star on this diagram is closely related to the star's 13) _____.

Exercise 18-3: Using Parallax*

In this exercise we will apply the concepts of parallax to making distance measurements on the Earth. These are commonly employed by surveyors to determine the distance to an inaccessible location such as the distance across a river.

Our goal in this exercise is to determine the distance to a yacht anchored far out from shore in Thunder Lake. A diagram illustrating this scenario can be found on the following page and will be developed in the following paragraphs. The major concept behind parallax is that nearby objects appear to be in different locations relative to more distant objects when the perspective of the observer changes. Thus, we want to look at the position of the yacht from two different locations separated

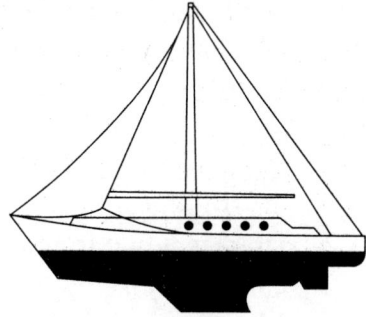

by a distance we call the baseline. The larger the baseline is, the more accurate the technique will be, but we also want the procedure to be convenient. Assume we have paced out or used a measuring tape to define a baseline of 320 meters (m).

We now want to make an angular measurement defining the position of the yacht from each end of the baseline. Surveyors have very accurate tools for making such measurements. They consist of a small spotting telescope attached to a precise protractor mounted on a tripod. Thus, the surveyor can sight on the yacht and measure the angle to the yacht relative to the baseline. Let's assume you have access to such an apparatus. From position A you measure an angle of 81° from the baseline to the yacht and from position B, 72°. You could do this very crudely with just a protractor.

We would like to determine the distance from the yacht to the baseline along a line perpendicular to the baseline. One method of doing this would be to approach the problem mathematically and apply trigonometry. To avoid mathematical complexity, we will take the simpler route of drawing a scaled diagram of the real-world situation on graph paper. In this diagram all distances will be proportional to the their real-world counterparts-only smaller. We first need to determine an appropriate scale so that our diagram will fit on the graph paper. The choice made here was to let the 320 m baseline be 16 blocks on the graph paper so that the scale is 1 block = 20 m. From position A we then draw a line at an angle of 81°. The intersection of this line with the line from position B defines the position of the yacht. We have now completed the diagram found on this page. You should now finish this parallax procedure and determine the distance to the yacht.

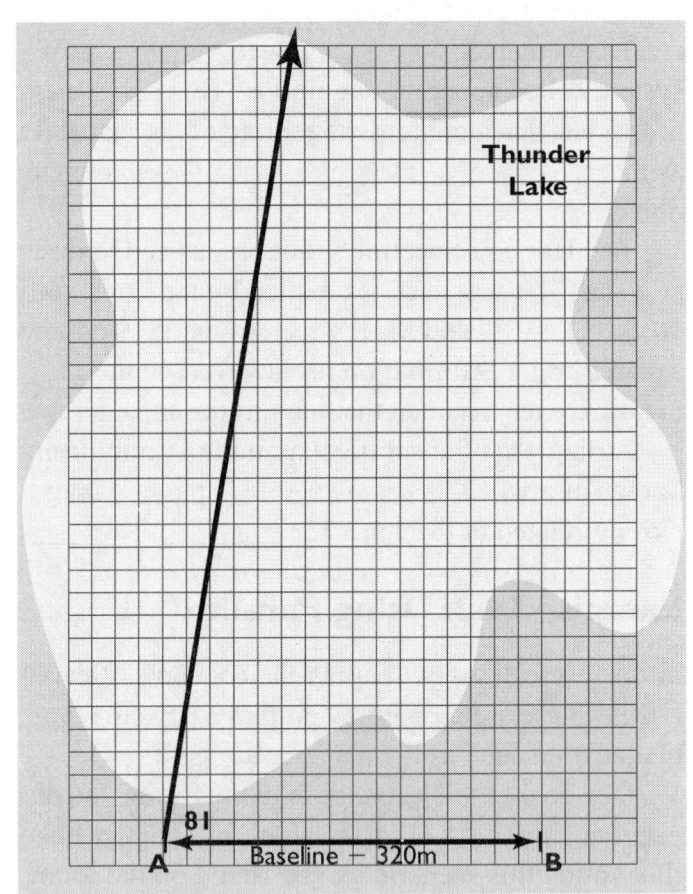

1. Draw in the line from position B at an angle of 72° with respect to the baseline. Draw in the yacht at the intersection of the two lines.
2. Draw in a line from the yacht to the baseline that makes an angle of 90° with the baseline.
3. Measure the length of this line:
   Length = _____ blocks
4. Convert this length into meters using the scale factor.
   Distance to yacht = _____

Exercise 18-4: Using Magnitudes*

Astronomers typically use a magnitude scale to classify stars because of the wide range of brightnesses that they have. We can use the following equation to relate the magnitude difference between two stars to the ratio of their brightnesses:

$$m_2 - m_1 = 2.5 \log\left(\frac{b_1}{b_2}\right)$$

We can reformat this equation as

$$\frac{b_1}{b_2} = (2.5)^{(m_2 - m_1)}$$

Magnitude Difference $(m_2 - m_1)$	Brightness Ratio (b_1/b_2)
0	1
1	2.5
2	6.3
3	16
4	40
5	100
6	250
7	630
8	1,600
9	4,000
10	10,000
15	10^6
20	10^8
25	10^{10}

Thus, each difference of one magnitude corresponds to a factor of 2.5 in intensity. So a first magnitude star is ($2.5 \times 2.5 \times 2.5 =$) 16 times as bright as a fourth magnitude star. This means that your eye is detecting 16 times as many photons of light.

The magnitude system described above is referred to as apparent magnitude. Stars with lower apparent magnitude appear brighter to us but may not be intrinsically more luminous because we haven't yet taken into account their distances. We do this with absolute magnitude (M). This is the apparent magnitude (m) of a star moved to a distance of 10 parsecs (pc). Thus, we imagine that we can move all of the stars onto a sphere with a radius of 10 pc that surrounds us. With the distance factor effectively removed, the apparent (now absolute) magnitude would allow us to identify the intrinsically more luminous objects. We can relate the apparent and absolute magnitudes through the following equation:

$$m - M = -5 + 5 \log_{10} d$$

Magnitude Difference $(m - M)$	Distance $(d$ in pc$)$
0	10
1	16
2	25
3	40
4	63
5	100
6	160
7	250
8	400
9	630
10	10^3
15	10^4
20	10^5

where the quantity $m - M$ is known as the distance modulus. Knowing the distance modulus allows one to calculate the distance. The values for both equations have been tabulated to help avoid mathematical complexity.

Directions: Use the two tables above to answer the following questions concerning the newly created 89th constellation. Note that the apparent magnitudes are given. Consider only labeled stars.

1. The star that would appear brightest to you is _____.

2. The star that would appear faintest to you is _____.

3. The light from _____ is 16 times less intense than the light from δ pistolis.

4. The light from _____ is 100 times more intense than the light from κ pistolis.

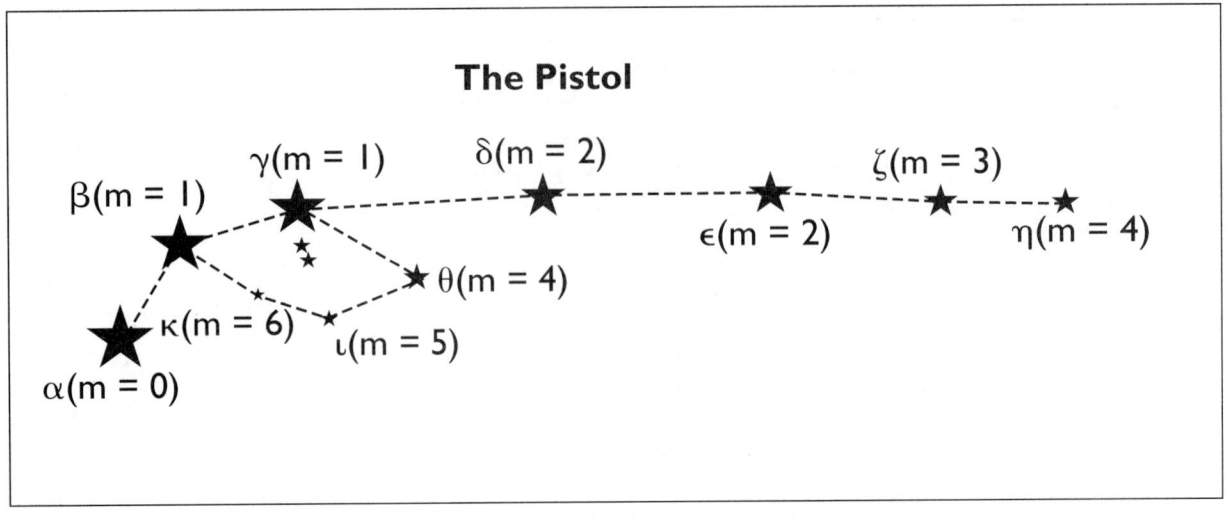

The Pistol

β(m = 1) γ(m = 1) δ(m = 2) ζ(m = 3)

α(m = 0) κ(m = 6) ι(m = 5) θ(m = 4) ε(m = 2) η(m = 4)

5. The intensity ratio between ζ pistolis and κ pistolis is _____.

6. The intensity ratio between β pistolis and ι pistolis is _____.

7. η pistolis is 10 parsecs distant; thus, its absolute magnitude M is

    _____.

8. γ pistolis is 40 parsecs distant; thus, its absolute magnitude M is

    _____.

9. The star ε pistolis has $M = -4$; thus, it is _____ parsecs distant.

10. The star β pistolis has $M = 0$; thus, it is _____ parsecs distant.

11. θ pistolis is 140 parsecs distant; thus, its absolute magnitude is _____.

12. The star δ pistolis has $M = -7$; thus, its distance modulus is _____.

Exercise 18-5: The HR Diagram*

Directions: Indicate the name(s) of the stars from the diagram to the right that correspond to each of the following statements.

1. The hottest star

    _____

2. The coolest star

    _____

3. The most luminous star

    _____

4. The least luminous star

    _____

5. The star with the largest radius

    _____

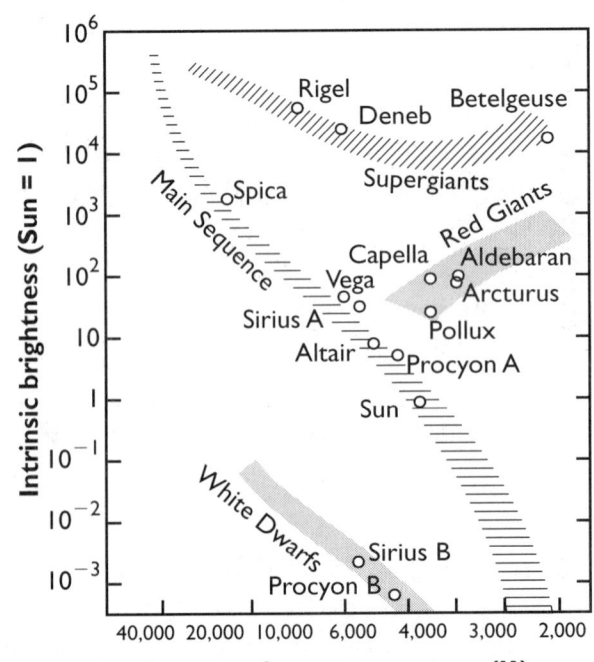

6. The star with the smallest radius

   _____

7. Main sequence stars

   _____

8. A spectral type AO star

   _____

9. A luminosity class la star

   _____

10. A star with titanium oxide lines in its spectra

   _____

Exercise 18-6: Spectroscopic Parallax*

In this assignment we will apply the distance determination technique known as spectroscopic parallax. To use it, an astronomer must obtain photometric observations of both the star's apparent magnitude and its spectra. You should keep in mind that even professional astronomers can achieve only 30% accuracy using Spectroscopic parallax. This exercise has a wide range of correct answers.

The procedure can be summarized as follows:

1. Use the particular absorption lines present in the spectrum to determine the spectral type from the chart below.

   Example: If your spectrum has strong hydrogen and weak ionized calcium lines (calcium is an ionized metal), you could classify it as approximately A0.

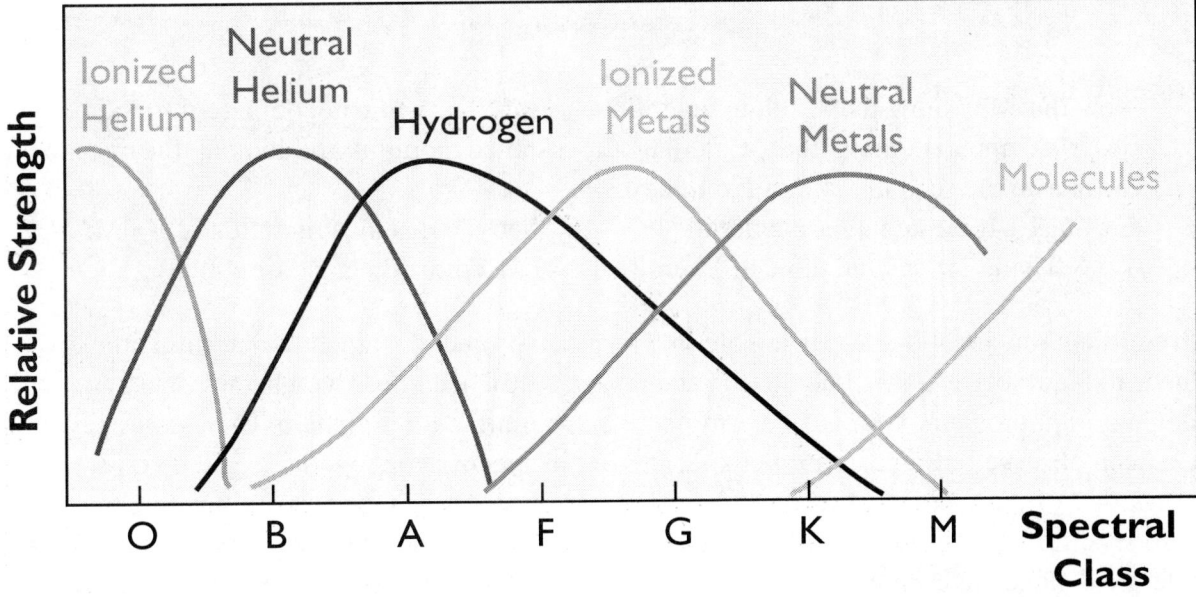

2. Use the thickness of the absorption lines present to determine the luminosity class from the diagram on the next page. Very thick lines will be luminosity class V (main sequence stars), and very thin lines will be luminosity type Ia.

Example: If your spectrum has very thick spectral lines, you could classify it as luminosity class V.

3. We now have uniquely specified the location of the star on the diagram below. Simply find the intersection of the spectral type (a column on the graph) and the luminosity class curve. Once you have noted this position, you can move horizontally to the left on the graph and read off the star's absolute magnitude.

Example: Find the intersection of spectral type A0 and luminosity class V on the diagram below. Now move horizontally to the left of this point and read off the absolute magnitude, which is about +1.0.

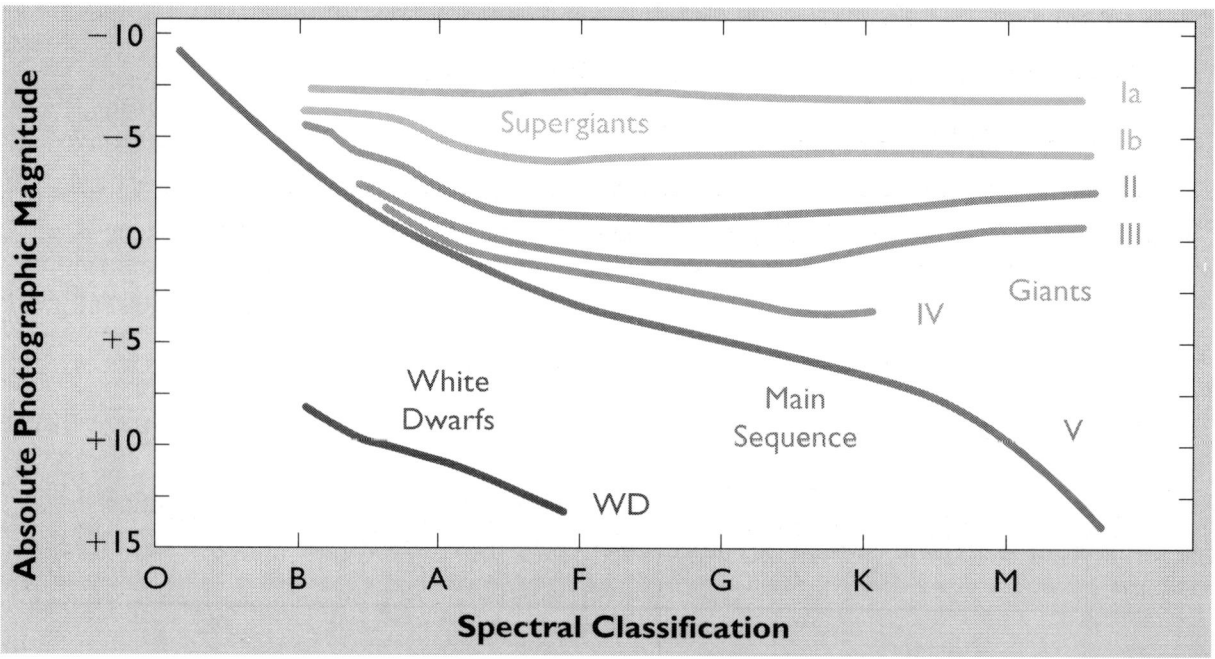

4. Now that you know the absolute magnitude of the star, you can compare it to the observed apparent magnitude. Calculate the distance modulus and look up the distance to the star in the second table of Exercise 18-4.

Example: If the apparent magnitude is $m = 9$, then the distance modulus is $m - M = 9 - 1 = 8$. Using the second table of Exercise 18-4, we find a distance of 400 pc.

Directions: You are about to participate in a research project designed to determine the distances to a number of stars. Each of the stars has already been observed by an astronomer specializing in photometry who has determined the apparent magnitude m of each star. They have also been observed by an astronomer specializing in spectroscopy who has given you a crude description of the types of absorption lines present in each star's spectra and the thicknesses of those spectral lines. Using the above two diagrams and the second table of Exercise 18-4, estimate the distance to each of the following stars using the technique of spectroscopic parallax. Complete each entry in the table on the following page as you go.

Star 1: $m_v = 9$, strong hydrogen and weak ionized calcium lines, very thick spectral lines

Spectral Type: _____   Luminosity Class: _____

Absolute Magnitude: _____   Distance Modulus: _____

Distance (in pc): _____

Star 2: $m_v = 4$, strong hydrogen and weak ionized calcium lines, very thick spectral lines

Spectral Type: _____   Luminosity Class: _____

Absolute Magnitude: _____   Distance Modulus: _____

Distance (in pc): _____

Star 3: $m_v = 6$, very weak hydrogen and very strong neutral iron lines, thin spectral lines

Spectral Type: _____   Luminosity Class: _____

Absolute Magnitude: _____   Distance Modulus: _____

Distance (in pc): _____

Star 4: $m_v = 7$, weak hydrogen and strong helium lines, very thick spectral lines

Spectral Type: _____   Luminosity Class: _____

Absolute Magnitude: _____   Distance Modulus: _____

Distance (in pc): _____

Star 5: $m_v = 10$, very weak hydrogen and titanium oxide lines, medium spectral lines

Spectral Type: _____   Luminosity Class: _____

Absolute Magnitude: _____   Distance Modulus: _____

Distance (in pc): _____

Star 6: $m_v = 8$, medium hydrogen and very strong ionized calcium, thick spectral lines

Spectral Type: _____   Luminosity Class: _____

Absolute Magnitude: _____   Distance Modulus: _____

Distance (in pc): _____

Exercise 18-7: True/False Questions

T / F   1. Energy production in a star comes from gravitational contraction and nuclear reactions in approximately equal parts.

T / F   2. In a fission reaction, two small nuclei are combined so that some of the mass is converted into energy.

T / F   3. The Coulomb barrier is the term for the electrical repulsion between two positively charged nuclei.

T / F   4. The Ideal Gas Law states that if volume is held constant, pressure will increase as temperature increases.

T / F   5. The CNO Cycle is a very important means of energy production in low-mass stars like the Sun.

T / F   6. The triple-alpha process can occur only at high temperatures because there are no stable nuclei with eight nucleons.

T / F   7. Degradation is the term used to describe the conversion of a high-energy photon produced in the core to a multitude of low-energy photons at the surface of the Sun.

T / F   8. Neutrinos produced in nuclear reactions make their way to the surface of a star in a random walk in exactly the same way that photons do.

T / F   9. Parallax measurements made from ground-based telescopes are just as accurate as those made from space-based telescopes.

T / F   10. If a star has zero proper motion, then it must be moving either directly toward us or directly away from us.

T / F   11. The frequency response of the eye is very similar to the B filter.

T / F   12. The difference between magnitudes in two different filters is called a color index and is useful for learning about the temperature of a star.

T / F   13. The primary reason for the different spectral types is that stars have different compositions.

T / F   14. Since white dwarfs have high surface temperatures and are faint, they must be very small.

Unit 19
Multiple Stars and Star Clusters

Chapter Objectives

Over half the stars in our sky are found in pairs (binary stars) or larger groups. Analysis of these binary systems provides us with our best means of determining stellar masses and sizes. In this chapter we will investigate the different types of binary stars and what can be learned from each of them about the physical properties of the stars. We will review Newton's generalizations of Kepler's laws as tools to determine stellar masses. The mass-luminosity relation that applies to most stars will be developed from these data on stellar masses. The Doppler effect will be applied to a study of binary star systems and also to the discovery of extrasolar planets. We will study the accretion of matter from one star onto another in some binary systems, and the relationship between that accretion and novae, supernovae, strong X-ray sources, and black holes.

The two types of star clusters—the younger open clusters and the older, more crowded globular clusters—will be introduced and the important information they reveal about stellar evolution will be discussed.

Progress Checklist

1. **Visual Binaries**
   - ❏ Visual Binaries
   - ❏ The Center of Mass
   - ❏ Astrometric Binaries
   - ❏ Masses for Binary Stars
   - ❏ Mass-Luminosity Relation
   - ❏ Planets and Binary Stars
2. **Spectroscopic Binaries**
   - ❏ Spectroscopic Binary Stars
   - ❏ Doppler Effect
   - ❏ Animation
   - ❏ Spectrum
   - ❏ Velocity Curves
   - ❏ Determining Masses
3. **Eclipsing Binaries**
   - ❏ Eclipsing Binaries
   - ❏ Algol Eclipsing Binary System
   - ❏ Eclipsing Animation
   - ❏ Stellar Diameters
4. **Accreting Binaries**
   - ❏ Accreting Binaries

   - ❏ Gravitational Potentials
   - ❏ Roche Lobes
   - ❏ Accretion Disks
   - ❏ Novae, Bursts, Supernovae
   - ❏ Black Holes
5. **Open Star Clusters**
   - ❏ Clusters and Groupings
   - ❏ Messier Objects
   - ❏ Open Clusters
   - ❏ Examples
   - ❏ Formation
   - ❏ Age and Evolution
6. **Globular Star Clusters**
   - ❏ Globular Clusters
   - ❏ Examples
   - ❏ Formation
   - ❏ Ages and Evolution
   - ❏ Stellar Populations
   - ❏ The Core

Keywords

visual binaries

center of mass

astrometric binary

Kepler's laws

Kepler's third law

tilt angle

proper motion

mass-luminosity relation

extrasolar planets

spectroscopic binaries

Doppler effect

redshift

blueshift

double-line spectroscopic
    binary

single-line spectroscopic
    binary

velocity curve

eclipsing binaries

light curve

Algol eclipsing binary

accreting binaries

angular momentum

accretion disk

X-ray binary

Roche lobes

inner Lagrange point

nova

supernova, type I

X-ray burster

black hole

open (galactic) cluster

globular cluster

associations

Messier objects

Pleiades cluster

Hyades cluster

Eagle nebula

Orion nebula

turnoff point

Exercise 19-1: Introductory Narrative

Most stars are not single stars like our Sun, but instead are part of two-star systems called binaries. Astronomers recognize several different types of binary stars. Systems where we can visually distinguish the two stars in orbit around each other are called 1) _____ binaries. If only one of the two stars is bright enough to be seen, astronomers may still be able to discern that it is a binary by the wobbles in its proper motion about the center of mass. Binaries that are spotted in this way are known as 2) _____ binaries. Binaries that are identified through Doppler shifts in their spectral lines are known as 3) _____ binaries. Graphs of radial velocity versus time are known as 4) _____ and provide detailed information about the system. Binaries where one star actually passes in front of the other along our line of sight are known as 5) _____ binaries. These systems must have an angle of inclination *i* near 6) _____. These systems are normally studied by graphing the variations in brightness as a function of time, which is called a(n) 7) _____. They afford astronomers one of the few direct ways by which stellar 8) _____ can be measured. Some binary pairs are close enough that matter from one star forms a(n) 9) _____ around the second star as the matter spirals into the second star.

   Stars are also found in larger groups known as clusters. 10) _____ clusters contain small numbers of young stars and are found within the disk of our galaxy. The stars they contain are high in metals and are known as population I stars. 11) _____ _____ clusters contain large numbers of small older stars and are symmetrically distributed about the center of our galaxy. The stars they contain are especially low in "metals" and are known as 12) _____ stars. This second type of cluster tends to be larger and contains about 1000 times as many stars as the first type.

Exercise 19-2: Eclipsing Binary Light Curves*

Directions: Complete the graphing and questions concerning the light curve of the eclipsing binary star system shown below. Note that star A has a hotter surface temperature than star B.

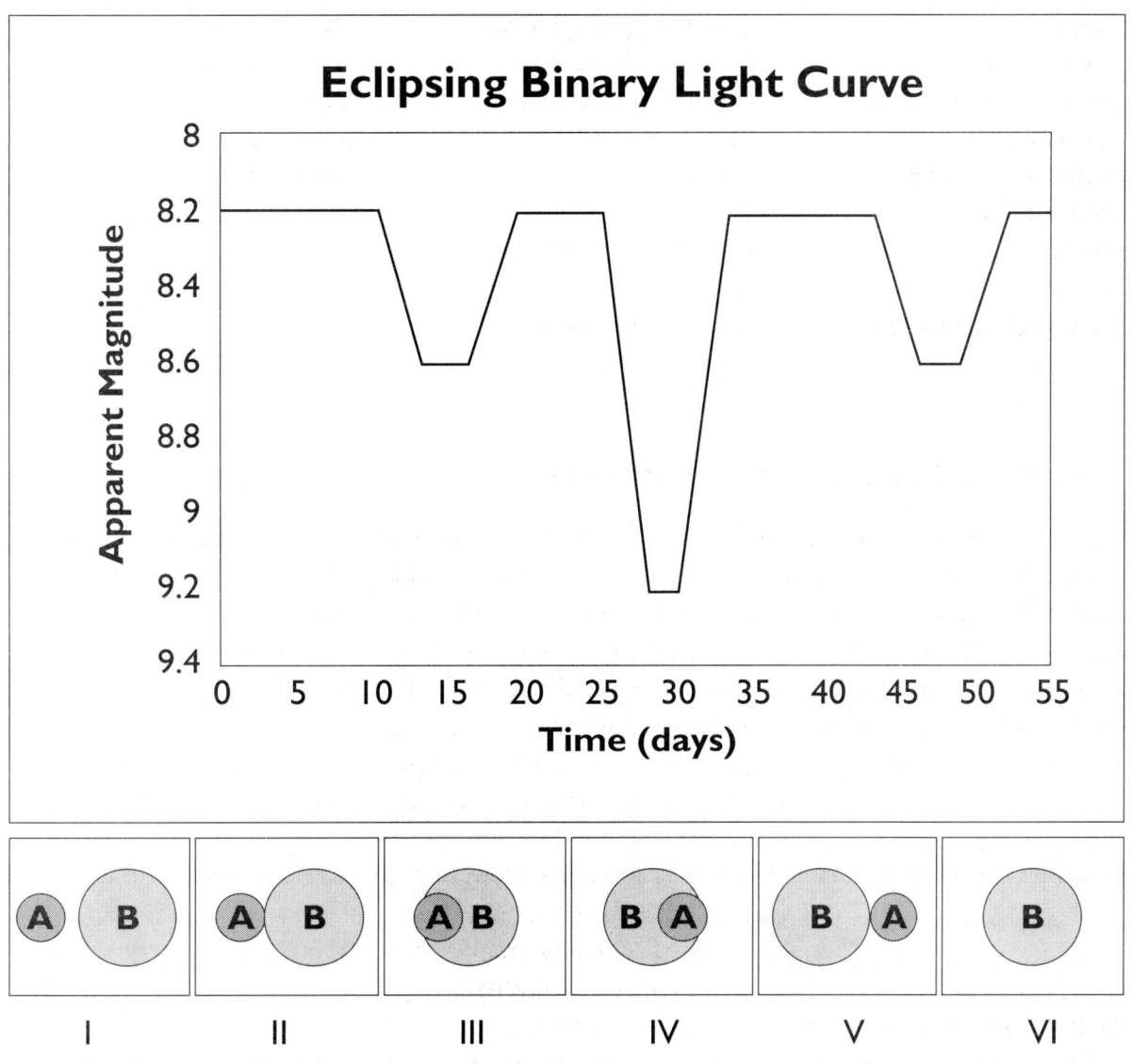

1. The six boxes above all correspond to a certain location on the light curve. Place the roman numeral corresponding to each figure at an appropriate location (time) on the light curve above.
2. The orbital period of the binary is _____.
3. The depth of the primary eclipse is _____, and the depth of the secondary eclipse is _____.

Exercise 19-3: Spectroscopic Binary Stars*

Directions: Complete the graphing and questions concerning the radial velocity curve of the double-line spectroscopic binary shown below.

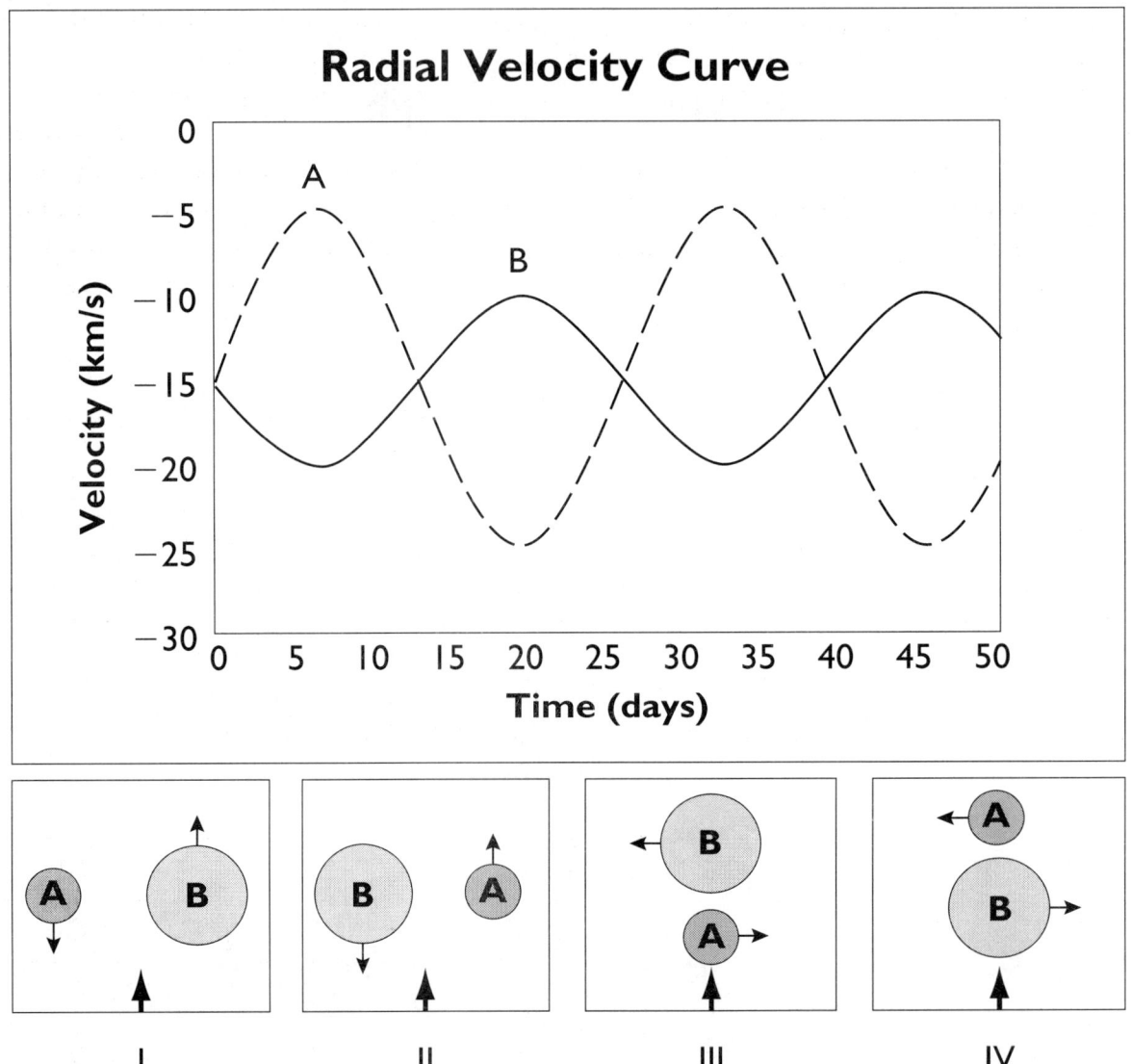

1. The four boxes above all correspond to a certain location on the radial velocity curve. The thick arrow at the bottom of each box represents the line of sight from the Earth, and the thin arrows represent the instantaneous velocities of the stars. Place the roman numeral corresponding to each figure at an appropriate location (time) on the radial velocity curve above.

2. The orbital period of this binary is _____.

3. The radial velocity of the center of mass of the system is _____.

4. Star _____ is the more massive of the two stars.

Exercise 19-4: The Ages of Open Clusters*

In this exercise we will use the turnoff point of a cluster on an HR Diagram to crudely estimate the age of the cluster. The turnoff point is the location on the HR Diagram where stars are starting to leave the main sequence. These stars are running out of fuel in their cores and are starting to move up the red giant branch. Since we know how long a star with a certain mass lives on the main sequence and we know the masses of stars at the various spectral types/surface temperatures along the main sequence, we can relate the two parameters.

The table entitled Main Sequence Spectral Class Properties from the *Properties of Stars/HR Diagram/Main Sequence* module is shown below. If we can identify the temperature (or spectral type) of the turnoff point on a cluster HR Diagram, we can use the chart to look up the corresponding age of the star just leaving the main sequence. Since all of the stars of the cluster formed at nearly the same time, this is also the age of the cluster. Since the table does not provide the time on the main sequence for every surface temperature, it will be necessary to estimate the age by interpolating between two given values.

Main Sequence Spectral Class Properties					
Spectral Class	Mass (Solar Units)	Luminosity (Solar Units)	Temperature (K)	Radius (Solar Units)	Time on Main Sequence (Million Years)
O5	40	400,000	40,000	13	1.0
B0	15	13,000	28,000	4.9	11
A0	3.5	80	10,000	3.0	440
F0	1.7	6.4	7500	1.5	2700
G0	1.1	1.4	6000	1.1	8000
K0	0.8	0.46	5000	0.9	17,000
M0	0.5	0.08	3500	0.8	56,000

Directions: Determine the turnoff point for each cluster shown on the following page. Be careful not to be fooled by any blue stragglers. Draw a vertical line at the temperature corresponding to the turnoff point and read off the temperature. Look up the temperature in the chart above and use it to estimate the age of the cluster in millions of years. Write in your estimated age below each cluster HR Diagram and circle any stars that you determine to be blue stragglers.

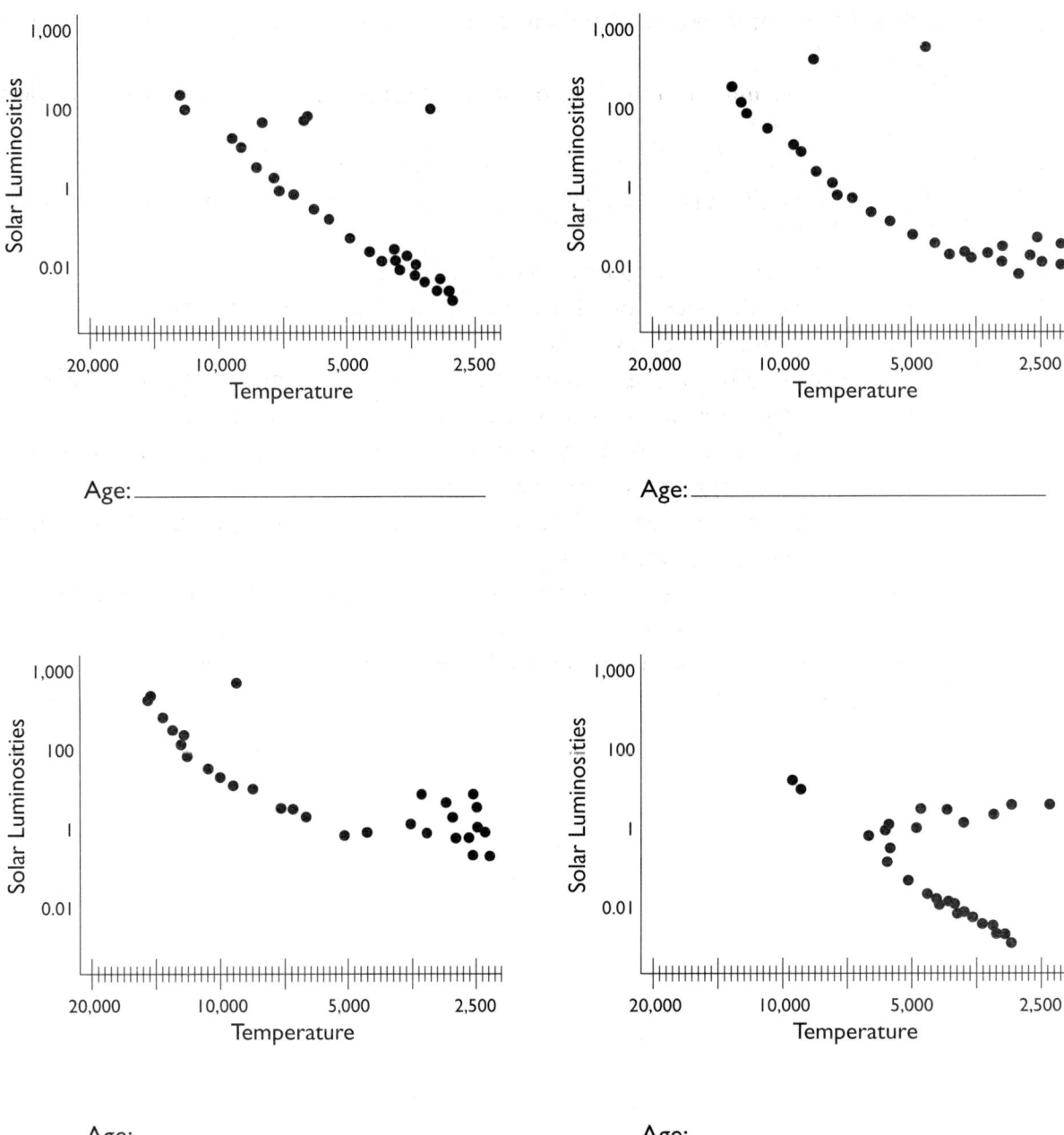

Age:_____

Age:_____

Age:_____

Age:_____

Exercise 19-5: True/False Questions

T / F   1. In the binary system Sinus A and B, the high-mass star Sinus A is stationary while the low-mass star Sinus B orbits around it.

T / F   2. The inclination of a binary's plane of orbital revolution is typically not known.

T / F   3. The luminosity of a 12 solar mass red giant could easily be calculated with the mass-luminosity relation.

T / F   4. It is often possible to determine the masses of the stars in astrometric binaries.

T / F   5. A spectroscopic binary where the spectral lines of one star are too faint to be seen is known as a single-line spectroscopic binary.

T / F   6. When one star of a spectroscopic binary is moving away from us, its spectral lines are blueshifted.

T / F   7. The inclination angle of the plane of orbit for eclipsing binaries can have any value between 0° and 90°.

T / F   8. Since the two stars in the Algol system are roughly the same size, the deepest eclipse in the light curve occurs when the spectral type B8 star eclipses the spectral type K2 star.

T / F   9. Accretion disks are often powerful sources of X-rays.

T / F   10. A nova involves the accretion of material from a star onto a neutron star where it ignites in a thermonuclear explosion.

T / F   11. The presence of a black hole is the best explanation for a binary system where there is a strong X-ray source and one star is massive but cannot be seen.

T / F   12. Stars in an association likely formed from the same cloud since they appear to be moving as a group but are too far apart to be bound by gravity.

T / F   13. Messier objects are galaxies, nebulae, and clusters that could possibly be mistaken for comets through a small telescope.

T / F   14. The stars in globular clusters must be old because they contain a high percentage of metals that hadn't yet been used up when the galaxy was young.

T / F   15. Harlow Shapley used the distribution of globular clusters to estimate the location of the center of our galaxy.

Unit 20
Star Birth and the Main Sequence

Chapter Objectives

In this chapter we will apply the basic physical principles of gravitational collapse to the nebulae where star birth occurs. The criteria necessary to begin the star formation process will be discussed, and stages in the life of a protostar will be illustrated. The most recent infrared images revealing these previously hidden star birth milestones will be presented. The evolutionary tracks calculated for forming stars will be traced out on the H-R diagram. The limiting masses for stars will be discussed and the brown dwarfs ("stars that missed") will be described.

Progress Checklist

1. Recipe for Stars
❑ Molecular Clouds
❑ Temperature, Pressure & Gravity
❑ Hydrostatic Equilibrium
❑ Jeans Collapse Criterion
❑ Fragmentation
❑ Sources of Instability

2. Protostars
❑ Cocoons for Young Stars
❑ Accretion Disks & Bipolar Outflows
❑ T-Tauri Stars

❑ Herbig-Haro Objects
❑ Motion on the HR Diagram
❑ Kelvin-Helmholtz Timescale

3. The Main Sequence
❑ Main Sequence Life
❑ Width of Main Sequence
❑ Dynamical Timescales
❑ Lower Mass Limit
❑ Brown Dwarfs
❑ Upper Mass Limit

Keywords

molecular clouds
molecular hydrogen
hydrostatic equilibrium
hydrodynamics
Jeans mass
Jeans density
fragmentation of cloud
spiral density waves
shock waves
protostars
cocoon

EGGs
accretion disks
bipolar flow
T-Tauri stars
Herbig-Haro objects
ejected jets
evolutionary track
Hyashi track
radiative core
Kelvin-Helmholtz timescale
virial theorem

main sequence lifetimes
ZAMS
dynamical timescales
free-fall timescale
expansion timescale
limiting lower mass
brown dwarfs
upper mass limit
Eddington luminosity
Wolf-Rayet stars

Exercise 20-1: Introductory Narrative

Stars are formed in giant clouds of gas and dust. Because it is cold enough in these clouds for two hydrogen atoms to be bound together, they are known as 1) _____.
The cloud can collapse if it has a mass greater than the 2) _____ mass, which is proportional to the temperature and radius and inversely proportional to the average mass of gas particles in the cloud. This collapse may be started by density waves in spiral galaxies or the 3) _____ from nearby supernovae. The cloud will typically break up into smaller collapsing clouds in a process known as 4) _____. As a cloud collapses, it heats up as a result of the conservation of 5) _____.
This increase in temperature is accompanied by an increase in the average velocity of the particles that inhibits further collapse. Thus, the rate of collapse is related to the rate of 6) _____ to the surface.

It is often difficult to study protostars because they are surrounded by gas and dust, which form a(n) 7) _____ around them. In the final stages of collapse, a protostar will be surrounded by an accreting disk of material and have jets of material known as 8) _____ streaming out along the axis of rotation.

It is thought that the Sun took about 9) _____ years to collapse to the main sequence. The path it traced out moving toward the main sequence on the HR Diagram is known as a(n) 10) _____. All main sequence stars are producing energy by fusing 11) _____ into helium. The most important quantity in determining a star's life is its mass. Stars on the main sequence have masses ranging from 12) _____ to over 100 solar masses. Both the time of collapse and the time spent on the main sequence decrease as the mass of the star increases.

Exercise 20-2: Luminosities and Lifetimes of Main Sequence Stars*

In this exercise we will calculate the luminosity and lifetime of a main sequence star of a given mass. Although we have been using these values for the last two units, here we will see where they actually come from. There are fairly simple formulas for both that make use of solar units, but a calculator will be necessary to complete this exercise.

   The luminosity of a star can be calculated using the mass-luminosity relation: $L = M^{3.5}$. We can use this to derive an expression for the lifetime, T, of a star on the main sequence. We know that the energy a star produces comes from nuclear reactions. Over the course of a star's lifetime, some fraction, f of its mass gets converted to energy. The process is governed by Einstein's equation $E = mc^2$. Thus, we can write an equation that relates all of the energy that a star produces in its lifetime to the total amount of mass converted to energy.

   Rather than estimate the quantity f, let's just assume that it is roughly the same for all stars. So, if we take our expression for stellar lifetime and divide by the same expression for the Sun, the quantity f cancels out. We are left with a simple expression in solar units.

$$LT = fMc^2$$

$$T = \frac{fMc^2}{L} = \frac{fMc^2}{M^{3.5}}$$

$$T = \frac{fc^2}{M^{2.5}}$$

$$\frac{T}{T_\odot} = \frac{\dfrac{fc^2}{M^{2.5}}}{\dfrac{fc^2}{M_\odot^{2.5}}}$$

$$T = \frac{1}{M^{2.5}}$$

Example: Let's apply these formulas to Sirius, the brightest star in our sky. Sirius has a mass of 2.3 solar masses. The luminosity is given by:

$$L = (M)^{3.5} = (2.3M_\odot)^{3.5} = 18.5L_\odot$$

The lifetime is given by:

$$T = \frac{1}{M^{2.5}} = \frac{1}{(2.3M_\odot)^{2.5}} = 0.12T_\odot$$

The estimated main sequence lifetime of the Sun is 10 billion years, so we can easily convert stellar lifetimes to actual years:

$$0.12T_\odot = 0.12T_\odot\left(\frac{1 \times 10^{10} \ years}{1T_\odot}\right) = 1.2 \times 10^9 \ years$$

   From this example we can see some general trends in stars. If a star is more massive than the Sun, it will be considerably more luminous. This occurs because the weight of the star pushing down on the core produces much higher core temperatures; thus, nuclear reactions occur at a faster rate. This causes the fuel in the core to be used up more rapidly, however, so more massive stars don't live nearly as long as the Sun.

Directions: Calculate the luminosity (in solar luminosities) and the main sequence lifetimes (in years) for the following stars.

Star #1: Proxima Centauri $M = 0.1M_\odot$

Luminosity

Main Sequence Lifetime

Star #2: Rigel $M = 10M_\odot$

Luminosity

Main Sequence Lifetime

You can chech your answers using the Calculator for the lifetime on the Main Sequence vs. Mass (**IC 13.1**)

Exercise 20-3: True/False Questions

T / F   1. Since UV radiation easily breaks molecules apart, giant molecular clouds must have considerable amounts of dust in them to shield the molecules from UV radiation from nearby stars.

T / F   2. As a cloud of gas contracts, its temperature will fall as gravitational potential energy is converted into kinetic energy.

T / F   3. As a star contracts, it is in hydrostatic equilibrium.

T / F   4. The Jeans mass increases with the temperature and radius of a cloud of gas.

T / F   5. A large collapsing nebula is likely to fragment into smaller collapsing clouds due to fluctuations that cause subregions of the nebula to exceed the Jeans density.

T / F   6. Shock waves from nearby supernovae may start the collapse of molecular clouds.

T / F   7. Evaporating gaseous globules are dense regions of a nebula where the material has been concentrated due to UV radiation from nearby stars.

T / F   8. Herbig-Haro objects are created when bipolar flows collide with the gas of the interstellar medium.

T / F   9. Collapsing protostars transport energy entirely by radiation.

T / F   10. The Kelvin-Helmholtz timescale states that massive stars take much longer than smaller stars to collapse to the main sequence.

T / F   11. The most important parameter for describing a star's life is its mass.

T / F   12. The width of the main sequence on the HR Diagram is due to differences in stellar masses.

T / F   13. The upper limit for the mass of a star is determined by radiation pressure because it increases much more rapidly with temperature than gas pressure.

T / F   14. The presence of lithium is useful in distinguishing between faint stars and brown dwarfs because lithium is produced in a star's fusion reaction.

T / F   15. Low-mass stars are difficult to find because they are faint and produce most of their radiation in the infrared.

Unit 21
Star Death

Chapter Objectives

What happens when a star runs out of hydrogen fuel in its core and reaches the end of its main sequence lifetime? In this chapter we will explore the stages a star evolves through as it approaches its inevitable death. The three end stages of stars (white dwarf, neutron star, and black hole) will be introduced and the processes leading to their production described. The red giant phase of life for a star like our Sun will be illustrated and the horizontal and asymptotic branches of red giant stars will be described. The types of variable stars that result from the pulsation in size that often accompanies the evolving states of giant and supergiant stars will be delineated. The nuclear processes that result in nucleosynthesis of heavy elements in dying stars will be described.

Progress Checklist

1. End of Main Sequence Life
- ❏ Lifetime on Main Sequence
- ❏ Three Endgames
- ❏ Binary Mass Transfer
- ❏ Evolution from Main Sequence
- ❏ Hydrogen Shell Burning
- ❏ Advanced Shell Burning

2. Red Giants
- ❏ Red Giant Evolution
- ❏ Red Giant Branch
- ❏ Thermonuclear Runaways
- ❏ Helium Flash
- ❏ Horizontal Branch
- ❏ Asymptotic Giant Branch

3. Planetary Nebula
- ❏ Examples
- ❏ Mass Loss in Red Giants
- ❏ Ejection of Envelope
- ❏ Track on HR Diagram

4. White Dwarfs
- ❏ White Dwarfs
- ❏ Size and Density
- ❏ Chandrasekhar Mass
- ❏ Novae

5. Variable Stars
- ❏ Variable Stars
- ❏ Cepheid Variables
- ❏ RR Lyrae Variables
- ❏ Pulsation Timescales
- ❏ Long-Period Variables
- ❏ Instability Strip

6. Supernovae
- ❏ Supernovae
- ❏ Type Ia Supernovae
- ❏ Type II Supernovae
- ❏ Supernova 1987A
- ❏ Supernova Candidates
- ❏ Supernova Remnants

7. Heavy Element Production
- ❏ Elemental Abundances
- ❏ Production of Light Elements
- ❏ Elements up to Silicon
- ❏ The Iron Peak
- ❏ The s-Process
- ❏ The r-Process

Keywords

main sequence lifetime	superwind	neutron star
red giant	Chandrasekhar mass limit	black hole
white dwarf	nova	nuclear density
planetary nebula	Cepheid variables	progenitor star
horizontal branch	RR-Lyrae variables	supernova remnants
asymptotic giant branch	pulsation timescales	nucleosynthesis
triple alpha process	hydrodynamical response	iron peak
electron degeneracy	times	radiative capture
fermions	long period red variables	s-process
bosons	instability strip	r-process
exclusion principle	supergiant	neutron capture
thermonuclear runaway	supernova, type I, type II	beta decay
helium flash	gravitational core collapse	chart of the nuclides
helium burning shell	shock wave	beta stability valley

Exercise 21-1: Introductory Narrative

Our Sun will live for about 1) _____ years on the main sequence burning hydrogen into helium. When the Sun has exhausted the hydrogen in its core, it will move toward the 2) _____ of the HR Diagram. Its energy will now come from 3) _____ burning in a shell surrounding the helium ash core, and since the energy source is now closer to the surface, the Sun will expand and become a(n) 4) _____. The helium core has no energy source to support the weight of the upper layers of the Sun, so it will contract and its temperature will rise. When the temperature of the now degenerate core reaches about 100,000,000 K, the triple-alpha process will vigorously begin in what is known as the 5) _____. Since the primary energy source is now back in the core, the Sun will contract and move back toward the main sequence. The Sun will now stably burn helium in its core in a location of the HR Diagram known as the 6) _____. The Sun will burn helium in its core (~7% of its life) for a much shorter time than it will burn hydrogen (~93% of its life). When the Sun runs out of helium in its core, it will again become a(n) 7) _____ as it burns helium in a shell surrounding the core of carbon ash. The Sun has insufficient mass to ever be hot enough to fuse carbon. The Sun will now become so large that it will blow off most of its outer layers into space in a phenomenon known as a(n) 8) _____. This exposes the hot inner layers of the Sun, causing it to move immediately to the far 9) _____ of the HR Diagram. The Sun cannot get any more energy from fusion and greatly resembles the dying embers of a fire. It will contract due to its own weight until gravity is halted by 10) _____. Then the Sun will slowly cool down and become a(n) 11) _____ about the size of the 12) _____.

Exercise 21-2: Using Pulsating Variable Stars as Distance Indicators*

Many of the later stages in a star's life can be very useful as distance indicators. All of the distance indicators we will use here are called "standard candles," which means that we have a good idea of how intrinsically bright they are-we know their absolute magnitude. We can observe their apparent magnitude and compare the two to get distance.

RR Lyrae stars are low-mass pulsating variables. Although the absolute magnitude of RR Lyrae stars varies slightly with pulsation period, one can see from the accompanying diagram that the variation is very small. For our purposes it is sufficient to assume that all RR Lyrae stars have absolute magnitudes of $M = +0.5$.

For the more massive Cepheid variables, we will need to observe the pulsation period of the star and then use it to obtain the absolute magnitude from the chart. It is also necessary to know the metallicity of the star since there is a considerable difference in the period-luminosity relation for Type I and Type II Cepheids.

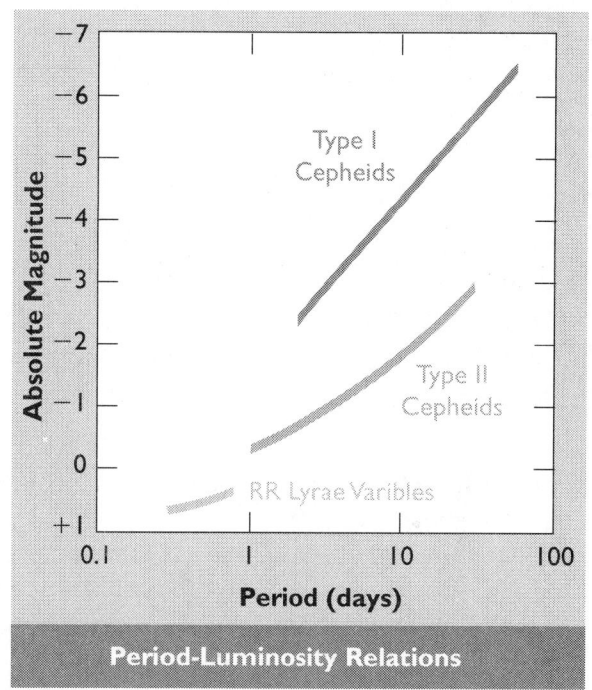

Period-Luminosity Relations

Example I: A Type I Cepheid variable is observed to have an apparent magnitude of $m = 6$ and a pulsation period of 50 days. How far away is the star?

Using the period-luminosity relation, we can see that a Type I Cepheid with a period of 50 days has an absolute magnitude of approximately -5. Thus, the distance modulus $m - M = 6 - (-5) = +11$. By extrapolating the pattern in the accompanying table, we can conclude that a distance modulus of 11 corresponds to a distance of 1600 parsecs (pc).

Distance Modulus $(m - M)$	Distance $(d$ in pc$)$
0	10
1	16
2	25
3	40
4	63
5	100
6	160
7	250
8	400
9	630
10	10^3
15	10^4
20	10^5

Example II: An RR Lyrae star has an apparent magnitude of 8. How far away is it?

$$m - M = 8 - (0.5) = +7.5$$

We can interpolate between the distance modulus values for 7 and 8 and get a distance of approximately 325 pc.

Directions: Estimate the distance to each of the following stars.

Star #1: An RR Lyrae star with $m = 9.5$

Distance

Star #2: A Type II Cepheid with a pulsation period of 5 days and $m = 4$

Distance

Star #3: A Type I Cepheid with a pulsation period of 10 days and $m = 6$

Distance

Exercise 21-3: True/False Questions

T / F 1. Stars leave the main sequence when approximately 10% of their original supply of hydrogen has been used up.

T / F 2. All of the red dwarfs that have ever formed are still on the main sequence because the universe is not yet old enough for them to have evolved off of it.

T / F 3. When the Sun runs out of hydrogen in its core, it will start burning helium.

T / F 4. The horizontal branch of the main sequence is a relatively stable region where core helium burning takes place.

T / F 5. Degenerate gases have the important property that their pressure is independent of temperature.

T / F 6. Planetary nebulae occur when the planets surrounding evolving stars are destroyed and the debris forms a shell around the star.

T / F 7. White dwarfs typically have hot surface temperatures due to the fusion of helium into carbon.

T / F 8. Chandrasekhar's limit states that all white dwarfs have to be more massive than 1.4 solar masses.

T / F 9. RR Lyrae stars are useful as distance indicators because they all have approximately the same intrinsic brightness.

T / F 10. Cepheid variables are useful as distance indicators because their luminosity is larger for stars with shorter pulsation periods.

T / F 11. The spectra from Type Ia supernovae do not show hydrogen lines because no hydrogen is present in the iron core.

T / F 12. The only difference between a Type Ia supernova scenario and a nova scenario is the rate at which matter accretes onto the surface of the white dwarf.

T / F 13. Astronomers were surprised to learn that the progenitor of Supernova 1987A was a blue supergiant instead of a red supergiant.

T / F 14. The nuclei with the highest nuclear binding energies are the light nuclei produced in the big bang.

T / F 15. The building of nuclei through the rapid capture of neutrons known as the r-process is thought to occur in Type Ia supernovae.

Conceptual Map 4
Stellar Evolution

In this exercise we will study the evolutionary tracks of stars on the Hertzsprung-Russell Diagram. In the following diagram, tracks are shown from the main sequence for stars of approximately 0.1, 1, and 20 solar masses. The evolutionary tracks for stars of a certain initial mass vary due to differences in composition and the amount of mass loss at various stages of their lives. Thus, these tracks only represent one possible evolution of the star.

First you should identify which track corresponds to which stellar mass. The terms below all correspond to a particular stage in one of the three stars' lives. Some of the terms apply to more than one star. Place the letter for each term at the appropriate location(s) on the appropriate evolutionary track(s).

A. Main Sequence Star
B. Helium Flash
C. Carbon Detonation
D. Red Giant
E. Horizontal Branch
F. White Dwarf
G. Planetary Nebula
H. Supernova
I. Fusion of Heavy Elements
J. Our Sun Today

If one could survey all of the many stars in the Milky Way, do you think we could find a star at virtually every point on the three evolutionary tracks? For which sections of which tracks would it be hard to find a star? Is there a part of one of the tracks for which no stars could be found in the Milky Way? Explain.

_____

_____

_____

_____

_____

_____

_____

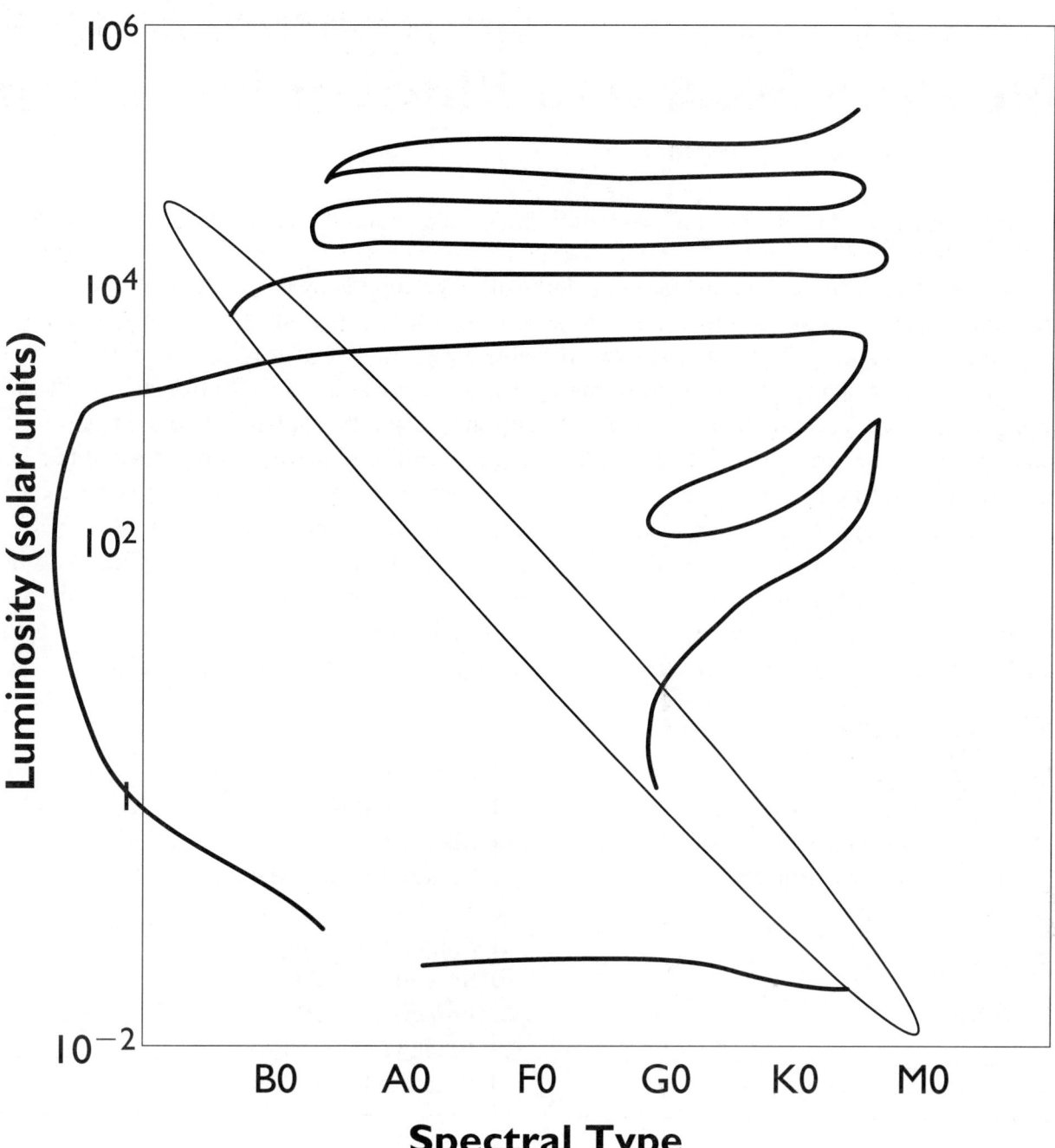

Unit 22
Neutron Stars and Black Holes

Chapter Objectives

At the end of their lives, the most massive stars will have collapsing cores that will exceed the Chandrasekhar limit and therefore these stars cannot leave white dwarfs as their stellar corpses. This chapter will examine the two possible core remnants that result when massive stars die: neutron stars and (for the most massive stars) black holes. The neutronization process that occurs during the formation of a neutron star will be explained and the basic structure of the neutron star that results will be described. The discovery of pulsars and the subsequent identification of these rapidly pulsing objects as rotating neutron stars will be chronicled. The upper limit to the mass of a neutron star will be shown to be about three solar masses; thus, a very massive star will undergo a core collapse that cannot be stopped, resulting in a black hole. The simple characteristics of a black hole will be described and the remarkable effects on the surrounding space-time will be illustrated. The search for stellar black holes and the list of the best candidates for black holes in binary star systems will be discussed.

Progress Checklist

1. Neutron Stars
- ❏ Neutron Stars
- ❏ Sizes and Densities
- ❏ Production of Neutron Stars
- ❏ Surface Gravity
- ❏ Magnetic Fields
- ❏ X-Ray Bursts

2. Pulsars
- ❏ Pulsars
- ❏ Lighthouse Model
- ❏ Rotation Rates

- ❏ Crab Pulsar
- ❏ Binary Pulsar
- ❏ Magnetars

3. Stellar Black Holes
- ❏ The Event Horizon
- ❏ Singularities, Naked & Clothed
- ❏ Spacetime Paths
- ❏ Properties of Black Holes
- ❏ Black Hole Candidates
- ❏ Rotating Black Holes

Keywords

neutron stars	binary pulsar	singularity
Chandrasekhar limit	general relativity	naked singularity
electron capture	curved spacetime	photon sphere
neutronization	gravitational waves	Einstein ring
pulsars	magnetars	no hair theorem
lighthouse model	soft gamma ray repeaters	Cygnus X-1
Crab pulsar	X-ray burster	ergosphere
synchrotron acceleration	event horizon	Kerr solution
starquakes	gravitational red shift	frame dragging
millisecond pulsars	Schwarzschild radius	

Exercise 22-1: Introductory Narrative

When low-mass stars like the Sun end their lives, they become 1) _____, but more massive stars become either neutron stars or black holes. Neutron stars are formed from stellar remnants more massive than 2) _____, up to possibly 3 solar masses although this upper limit is not well known. At these larger masses, electron pressure is insufficient to support the weight of the material, so electrons and 3) _____ are forced together to form neutrons and the object collapses to about the size of 4) _____.

Neutron stars have very powerful 5) _____ fields. Due to the conservation of angular momentum, neutron stars also rotate 6) _____. The combination of these two effects causes strong electric fields that accelerate electrons away from the surface near the magnetic poles, producing 7) _____ radiation in a pair of beams. Because the magnetic axis doesn't coincide with the rotation axis, these beams trace out a corkscrew pattern as the neutron star rotates. If one of the beams happens to cross our line of sight, we perceive the neutron star to be a(n) 8) _____. This explanation for the phenomenon is known as the 9) _____ model.

More massive stellar remnants become black holes. The matter collapses down to an object of zero size known as a(n) 10) _____. Gravity is so strong near a black hole that the escape velocity is equal to the speed of light. The region surrounding a black hole from which light cannot escape is known as the 11) _____. Black holes are extremely simple because all information concerning their constituent matter is destroyed as they form. Their only distinguishing characteristics are mass, 12) _____, and angular momentum.

Since light cannot escape from black holes, the only way to detect them is indirectly through the effects they have on surrounding matter. This typically involves detecting the 13) _____ that are produced from friction as material spirals into a black hole.

Exercise 22-2: True/False Questions

T / F   1. An electron and a proton can combine to form a neutron and an electron neutrino.

T / F   2. Neutron stars can have masses between 0.8 and about 3 solar masses.

T / F   3. Neutron stars are thought to be the cores of imploded Type I supernovae.

T / F   4. Supernovae explosions are thought to be asymmetric due to the high space velocities of neutron stars.

T / F   5. Neutron stars have extremely high surface gravity and strong magnetic fields.

T / F   6. When matter accretes onto the surface of a neutron star from another star in a binary system, a nova may occur.

T / F   7. When first detected, the regularity of the signals from pulsars led some to suggest they were "little green men."

T / F   8. The magnetic axis of a neutron star is well aligned with its rotation axis.

T / F   9. As a pulsar gets older, its rate of spinning gradually increases as further gravitational collapse occurs.

T / F 10. The sudden increases in pulsar periods known as glitches are probably due to seismic events.

T / F 11. Millisecond pulsars spin extremely rapidly because they are very young and haven't had time to lose much energy.

T / F 12. The region surrounding a black hole for which the escape velocity is the speed of light is known as the event horizon.

T / F 13. Light can travel in a circular orbit around a black hole.

T / F 14. The statement "black holes have no hair" refers to the smoothness of the photon sphere.

T / F 15. Black holes are most often identified by X-rays produced by material spiraling into them.

Unit 23
The Milky Way

Chapter Objectives

The Universe is composed of billions of separate collections of stars known as galaxies, and our home galaxy is an example of the spiral type of galaxy. Our Sun is located in the outer reaches of the Milky Way Galaxy and is one of the several hundred billion stars that comprise this typical spiral. In this chapter we will study the parts of the Milky Way Galaxy: the halo, the disk, and the central bulge. The views of our galaxy in different wavelengths will be illustrated and the resulting evolution of our perception of our home galaxy will be discussed. The spiral structure found in the disk portion of our galaxy will be described and the mechanisms that may have caused this spiral pattern will be explored. The concept of dark matter and its role in the dynamics of our galaxy will be introduced.

Progress Checklist

1. Observing the Galaxy
- ❏ The Milky Way
- ❏ Our Evolving Perception
- ❏ Measuring the Galaxy
- ❏ Visible Light View
- ❏ Infrared View
- ❏ Radio Maps

2. Components of the Galaxy
- ❏ Components
- ❏ Galactic Disk
- ❏ Visible Halo
- ❏ Galactic Bulge
- ❏ Magnetic Field & Cosmic Rays
- ❏ Dark Matter Halo

3. Rotation and Spiral Structure
- ❏ Spiral Structure
- ❏ Density Waves
- ❏ Self-Sustaining Star Formation
- ❏ Rotation Curves & Galaxy's Mass

4. The Interstellar Medium
- ❏ The Interstellar Medium
- ❏ Interstellar Gas
- ❏ HI & HII Regions
- ❏ Interstellar Dust
- ❏ Nebulae
- ❏ Aluminum-26

Keywords

Milky Way	galactic latitude	disk
Milky Way galaxy	galactic poles	spiral arms
grindstone model	infrared telescope	spherical component
Cepheid variables	radio telescope	galactic magnetic field
spiral galaxy	neutral hydrogen	dark matter
kiloparsec	21 cm line	Population II
megaparsec	spin-flip transition	Population I
galactic coordinate system	central bulge	metal poor
galactic longitude	halo	metal rich

interstellar medium   differential rotation   interstellar dust grains
Sagittarius A*   grand design spirals   dark (absorption) nebulae
polarization of light   rotation curve   interstellar extinction
cosmic rays   emission nebulae   interstellar reddening
H-II regions   reflection nebulae   forbidden transitions
spiral tracers   carbon monoxide   metastable state
density waves   coronal gas of ISM   Aluminum-26
self-sustaining star formation   HI region

Exercise 23-1: Introductory Narrative

Our Solar System is one of approximately 1) _____ stars that along with gas and dust make up the Milky Way Galaxy. Since we are located inside the Milky Way, it was difficult for astronomers to discern the nature of the galaxy. Two major breakthroughs occurred in the 1920s using Cepheid variable stars as distance indicators. 2) _____ _____ determined the scale of the Milky Way. He mapped the location of globular clusters, and by assuming that they were distributed in a spherically symmetric pattern, he calculated the distance to the center of our galaxy. 3) _____ found that the distance to the Andromeda Galaxy is much larger than the size of our own galaxy. He showed that galaxies were really "island universes," huge isolated collections of stars.

The Milky Way has three visible components known as the disk, the halo, and the 4) _____. These components are a result of the way our galaxy formed from a large cloud of gas and dust. Originally, the cloud was spinning rather 5) _____ and had a(n) 6) _____ abundance of metals. The first stars to form were population 7) _____, spherically distributed, and in randomly oriented elliptical orbits. As the galaxy contracted due to gravity, it began to spin more 8) _____ and formed a disk where most of the gas and dust became concentrated. More recent stars that have formed there are population 9) _____ and travel in circular orbits in the plane of the disk of our galaxy.

We can learn about the distribution of matter in our galaxy by looking at the orbits of stars. The graph of the orbital velocities of stars versus their distance from the galactic center is called a(n) 10) _____. It shows that orbital velocities increase toward the outskirts of the Milky Way. For these stars to be orbiting so rapidly, their orbits must enclose a large amount of mass that astronomers don't see. Thus, most of the matter of our galaxy is unseen and is known as 11) _____.

Exercise 23-2: Components of the Milky Way Galaxy

Directions: Below is a list of facts and observations for 15 different stars. Indicate whether the data suggest that the star belongs to the disk (D) or the halo (H) of the Milky Way

_____   1.   This star contains 3% metals.

_____   2.   This star contains 0.4% metals.

_____   3.   This star is in the open cluster known as the Pleiades.

_____   4.   Doppler shifts indicate that this star has a high radial velocity.

_____   5.   Parallax is large for this star, yet the proper motion is very small.

_____   6.   This star is part of an association in the Orion spiral arm.

_____   7.   This star is an RR Lyrae star in the globular cluster M3.

_____   8.   This star has a very elliptical orbit around the galaxy.

_____   9.   This star is a Type II Cepheid variable.

_____   10.   This star is an O3 star.

_____   11.   This star formed when our galaxy had a spherical shape.

_____   12.   This star is far away and has an high galactic latitude.

_____   13.   This star was formed by "self-sustaining star formation."

_____   14.   This star is between a giant molecular cloud and an HII region.

_____   15.   The spectra of this star are weak in the lines of heavier atoms.

Exercise 23-3: True/False Questions

T / F   1. For observers in the northern hemisphere, the summer sky's Milky Way is brighter than the winter sky's Milky Way because it is in the direction of the center of the galaxy.

T / F   2. Galactic longitude is the angle between the plane of the disk of our galaxy and a line from our Solar System to the object of interest.

T / F   3. We can view the center of our galaxy better in infrared light than visual because the hot stars there produce more light in this wavelength band.

T / F   4. A population I star is likely to be much older than a population II star.

T / F   5. The halo of our galaxy has a diameter at least as large as the disk's diameter.

T / F   6. Halo stars have circular coplanar orbits.

T / F   7. Because elongated dust particles spin so that they are aligned with any magnetic fields that are present, scattered light is polarized and can be used to study the magnetic field.

T / F   8. The spiral arms of our galaxy are bright because of the supernovae that occur in them.

T / F   9. The definitive characteristic of a spiral tracer is that it must be short-lived so that it cannot move more than the width of a spiral arm in its lifetime.

T / F   10. It is difficult to explain flocculent spiral galaxies using the density wave theory for spiral arms.

T / F   11. The fact that rotational velocities increase at the edge of the Milky Way proves that many black holes are present there.

T / F   12. Although the interstellar medium typically has very low density, it makes up about 15% of the total mass of visible matter in the Milky Way.

T / F   13. Some clouds of gas are known as HII regions because two hydrogen atoms are loosely bound together.

T / F   14. Interstellar reddening occurs because the gas of the interstellar medium preferentially scatters blue light more than red light.

T / F   15. Forbidden transitions occur in nebulae because the density is so low that collisions between atoms are very infrequent.

Unit 24
Galaxies

Chapter Objectives

The visible Universe contains billions of galaxies that collect together in clusters, superclusters, bubbles, and "Great Walls" that are the large-scale structure of our Universe. In this chapter we will describe the three main classes of galaxies: spiral, elliptical, and irregular galaxies. The differences between the galaxy types will be illustrated and the suggested reasons behind these differences will be explored. The clusters and superclusters of galaxies will be described and the effects of galaxy interactions and collisions in the crowded clusters of galaxies will be shown. The large scale soap bubbles, voids, and walls that appear in three-dimensional structural "maps" of our Universe will be introduced.

Progress Checklist

1. Hubble Classification
- ❏ The Tuning Fork Diagram
- ❏ Elliptical Galaxies
- ❏ Spiral Galaxies
- ❏ Irregular Galaxies
- ❏ Summary of Galaxy Properties
- ❏ Galactic Evolution?

2. Clusters and Superclusters
- ❏ Groups of Galaxies
- ❏ Clusters of Galaxies
- ❏ Mass Contained in Clusters
- ❏ Dark Matter
- ❏ Superclusters
- ❏ The Great Attractor

3. The Expanding Universe
- ❏ Cosmic Distance Ladder
- ❏ Redshifts
- ❏ The Hubble Law

- ❏ Age of the Universe
- ❏ Look-Back Times
- ❏ The Most Distant Objects

4. Soap Bubbles and Voids
- ❏ Redshift Surveys: the 3D Universe
- ❏ Soap Bubbles and Voids
- ❏ Great Walls
- ❏ 3D Structure of Universe
- ❏ Computer Simulations
- ❏ Summary of Distance Scales

5. Interacting Galaxies
- ❏ Colliding Galaxies
- ❏ Starburst Galaxies
- ❏ Cosmic Cannibalism
- ❏ Interactions & Evolution
- ❏ The Early Universe
- ❏ Collision of Andromeda

Keywords

Hubble Classification	giant elliptical	groups of galaxies
tuning fork diagram	dwarf elliptical	The Local Group
elliptical galaxies	peculiar galaxies	Andromeda Galaxy
spiral galaxies	galactic cannibalism	Large Magellanic Cloud
barred spirals	clusters of galaxies	Small Magellanic Cloud
irregular galaxies	superclusters	Sagittarius Dwarf

rich clusters	nonrelativistic matter	Hubble Law
poor clusters	The Local Supercluster	Hubble constant
Virgo cluster	Hubble flow	Hubble time
Coma cluster	peculiar velocity	look-back times
dark matter	The Great Attractor	pencil surveys
X-ray gas	standard candle	voids
baryonic matter	Tulley-Fisher Relation	Great Wall
nonbaryonic matter	type Ia supernova	Southern Wall
supersymmetric particles	Olber's Paradox	colliding galaxies
hot dark matter	Cepheid variables	starburst galaxies
cold dark matter	galactic red shift	Cartwheel Galaxy
relativistic matter	red shift parameter	

Exercise 24-1: Introductory Narrative

Hubble classified galaxies morphologically into groups. He called two of the groups spiral and elliptical; those that didn't fit into either of the previous groups were known as 1) _____ _____ galaxies. The spiral galaxies can be further divided into normal spirals and those that have 2) _____ running through their centers. Both the spiral and the elliptical groups are then further subdivided by their shape. Elliptical galaxies are classified as E0 for those that are 3) _____ to E7 for highly flattened galaxies. Spirals are further classified as Sa (or SBa) for galaxies with 4) _____ nuclei and tightly wound spiral arms to Sc (or SBc) for galaxies with small nuclei and 5) _____ wound spiral arms. All of these types are organized into a structure known as the 6) _____.

Types of galaxies differ in other ways besides shape. Spiral galaxies have considerably more gas and 7) _____ than do elliptical galaxies. Since these are the materials needed for 8) _____, we see massive young stars that have formed very recently in spirals but not in ellipticals. Consequently, spirals are considerably brighter on average than elliptical galaxies. Thus, the majority of galaxies that we see are spirals even though ellipticals are really the most abundant type of galaxy. Another difference is that ellipticals are more commonly found in regions of space that are densely packed with galaxies. This suggests that galaxies 9) _____ due to interactions with their neighbors.

Galaxies are assembled on large scales in groups, clusters, and 10) _____. All of these are held together by the mutual 11) _____ attraction between the galaxies. However, only about 1/10 of the masses needed to keep the clusters of galaxies held together is emitting visible light, which indicates the presence of 12) _____.

Exercise 24-2: Distances to Galaxies

In this exercise we will use supernovae as standard candles to estimate the distance to remote galaxies. The ideas being used here are very similar to those in Exercise 21-2 where we used pulsating variable stars as distance indicators. Astronomers have a good idea what the absolute magnitude of a supernova is at its peak brightness. They can then compare this peak absolute magnitude with the observed peak apparent magnitude and calculate the distance. Because supernovae are extremely luminous, this technique can be used on galaxies over 1000 Mpc away, whereas Cepheid variables are limited to about 20 Mpc.

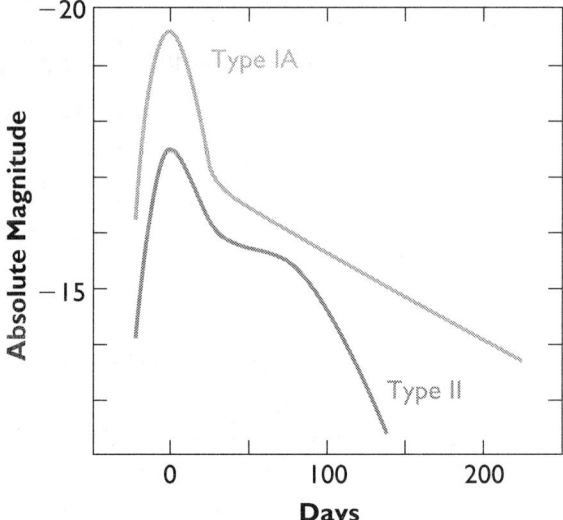

Typical Supernova Light Curves

Example: A Type II supernova is observed for several weeks, and its light curve has a peak apparent magnitude of 13.0. We know from the graph shown to the right that Type II supernovae peak with absolute magnitudes of approximately 17.5.

$$m - M = 13.0 - (-17.5) = 30.5$$

Using the chart to the right, one can see that 30.5 would be between 1.0×10^7 and 1.6×10^7. So let's call our final answer 1.4×10^7 or 14 Mpc.

Galaxy #1: A Type II supernovae is observed with a peak apparent magnitude of 20.5. Estimate the distance to this galaxy.

Distance =

Galaxy #2: A Type IA supernovae is observed with a peak apparent magnitude of 14.3. Estimate the distance to this galaxy.

Distance =

Distance Modulus ($m - M$)	Distance (d in pc)
0	10
1	16
2	25
3	40
4	63
5	100
6	160
7	250
8	400
9	630
10	10^3
15	10^4
20	10^5
25	10^6
30	10^7
35	10^8
40	10^9

Exercise 24-3: Galaxy Classification

In this exercise we will look at actual pictures of galaxies and try to classify them according to the Hubble Tuning Fork Diagram. This classification breaks galaxies up into the large groups called spirals (which has 6 subclasses) and ellipticals (which has eight subclasses). There are two additional classifications: S0, which serves as a transition between the spiral and elliptical groups, and IRR (irregular galaxies), which is a catchall for galaxies that don't fit into any of the other groups.

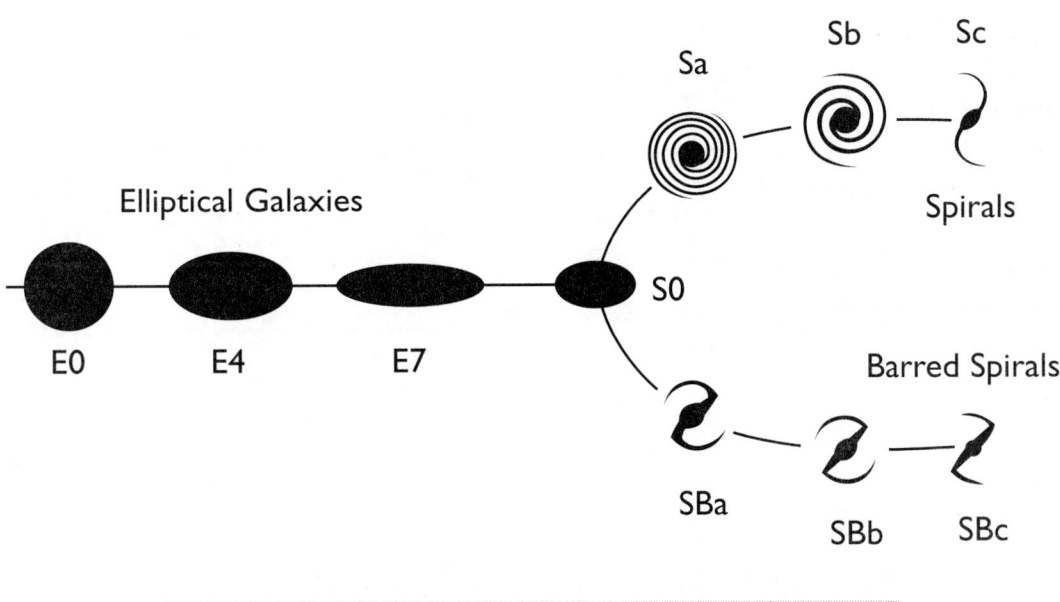

Hubble Classification of the Galaxies

The following flowchart illustrates the thought processes that should be used to accomplish this procedure. The diamond symbols indicate a "fork in the road" where a decision must be made concerning classification. One should first ask, "Is there a disk component?" If a disk is present, the galaxy is a spiral and must be further categorized based on the winding of the spiral arms, nucleus size, and whether a bar exists in the center of the galaxy. If no disk is present, one must next ask, "Is it spherical or elliptical in shape?" If the answer is yes, the galaxy is an elliptical and must be further classified based upon its exact shape. If no, then the galaxy must be an irregular.

Galaxy classification is not an exact procedure. There will always be some imprecision when deciding the exact classification. In addition, because galaxies are viewed in random orientations, it is occasionally difficult to determine even the basic group to which the galaxy belongs. For example, spiral galaxies that are viewed from a perspective nearly in the plane of the disk are particularly troublesome. Since one cannot see the spiral arms, classification relies more heavily on the size of the nucleus and whether gas and dust are visible.

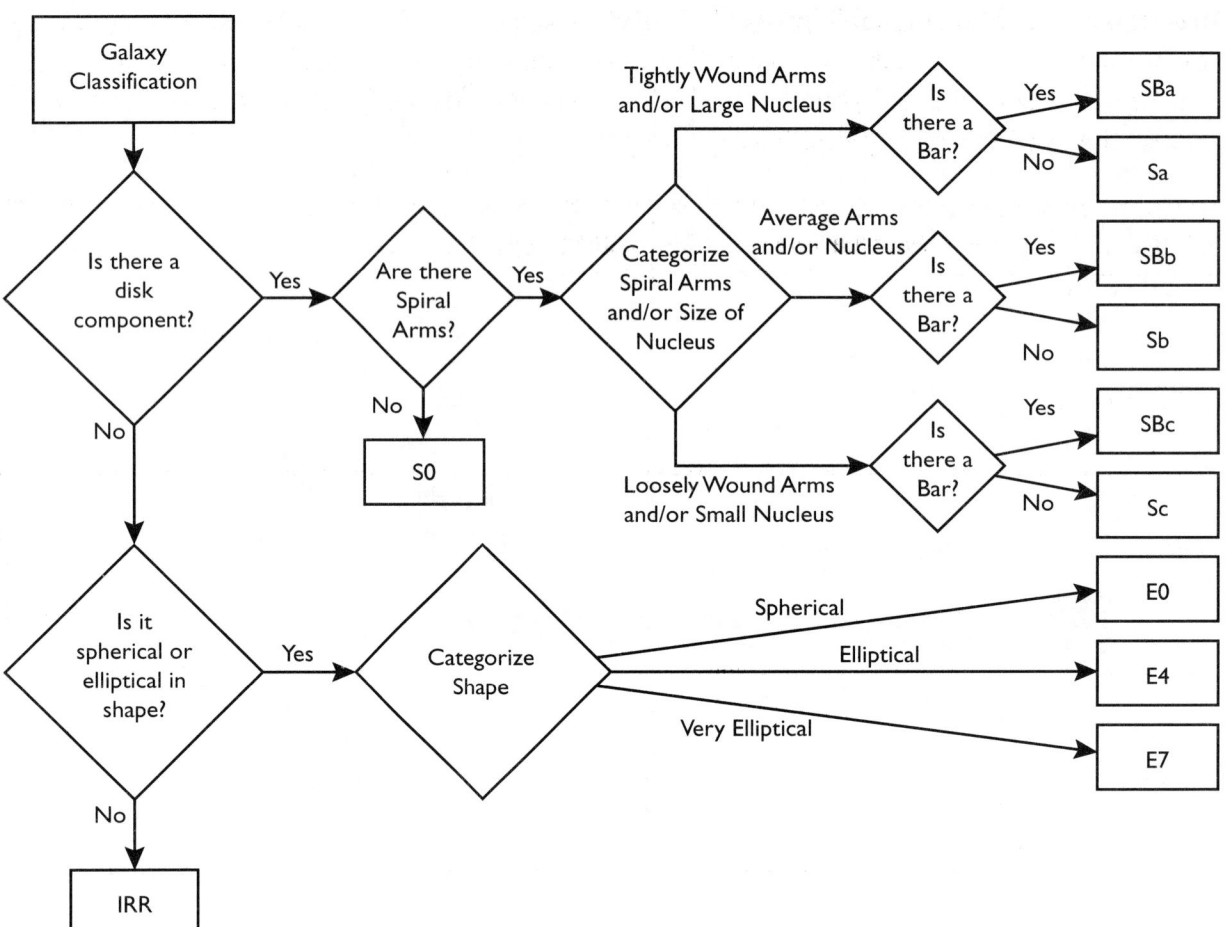

Directions: The following table gives URLs for images of a variety of galaxies. You should look at each galaxy and classify it according to Bubble's classification sequence. Then indicate some of the reasons that led you to your decision. The base of the URL is **http://www.seds.org/messier/jpg/** and is the same for each image.

URL	Classification	Justification
Base+ **m64.jpg**		
Base+ **m82.jpg**		
Base+ **m87.jpg**		
Base+ **m33.jpg**		
Base+ **m32.jpg**		
Base+ **m91.jpg**		

The known classifications and other information about these galaxies can also be found on the seds site. For example, for the first galaxy the URL would be **http://www.seds.org/messier/m/m064.html**

Exercise 24-4: True/False Questions

T / F   1. The Hubble Tuning Fork Diagram is useful because it shows how galaxies evolve from E0 to either Sc or SBc.

T / F   2. Elliptical galaxies have a far greater range of sizes than do spiral galaxies.

T / F   3. Elliptical galaxies typically have far more gas and dust than do spirals.

T / F   4. The spiral arms of a spiral galaxy are visible due to the large concentration of small faint stars found there.

T / F   5. Approximately 70% of the galaxies we see are spiral, so they must be the most common type of galaxy.

T / F   6. Since the Andromeda Galaxy is about 3 million light-years away from us, the most recent information we can obtain about it is 3 million years old.

T / F   7. It is difficult to determine the number of members in our Local Group of galaxies because the dust in the plane of the Milky Way may obscure some of them.

T / F   8. If a typical cluster of galaxies has a mass to light ratio of about 400, the cluster must contain many bright galaxies.

T / F   9. Because there are so many neutrinos in the Universe, if only one of the families of neutrinos had a nonzero mass, that could possibly be sufficient to answer the dark matter question.

T / F   10. An extremely massive object known as the Great Attractor causes a large-scale streaming of galaxies in the direction of Centaurus known as the Hubble flow.

T / F   11. Olber's Paradox or "Why is the night sky dark?" is resolved by the expansion of the universe.

T / F   12. One can estimate the age of the universe by calculating the reciprocal of the Hubble constant.

T / F   13. Redshift surveys of galaxies have shown structure on a scale even larger than that of superclusters.

T / F   14. In starburst galaxies the production rate of new stars may be as high as two to three per year.

Conceptual Map 5
The Cosmic Distance Ladder

In this assignment we will complete a logarithmic distance axis from one parsec to 1000 Mpc and describe the methods of distance determination that are used at various distances. This exercise will require you to tie together concepts from the last few chapters.

The general format of this Conceptual Map is a box with a small area on top to contain the name of the distance determination technique and a larger area below to describe exactly how the method is used. In each of these boxes either the name of the technique or the description are already completed for you and you are expected to add the other component.

Distance Determination Technique
Description of Technique

RADAR
In this technique, radio waves are transmitted toward a planet or asteroid. The time between transmission and reception of the returning echo is measured. The distance to the body (times 2) is then found by multiplying the elapsed time by the speed of light.

Our distance axis covers the Milky Way galaxy and the Universe. One could have extended the lower range of the distance axis into the solar system and discussed the determination of distances to the planets. An example box describing radar is shown to the right.

A line is drawn connecting the box to an appropriate distance on the axis where the technique might be applied. Since all commonly used techniques are valid at a range of distances, we will associate a technique with one of the larger distances at which it is typically used. Note that the ranges of the various techniques often overlap. For example there are RR Lyrae stars which are close enough so that parallax measurements may be made for them. Thus, the distances to the nearer RR Lyraes are "calibrated" using parallax measurements. Distances can then be determined to RR Lyraes that are too far away for parallax.

This is where the term "cosmic distance ladder" originates. Each method representing a rung of the ladder is used to refine the technique applied at greater distances which represents the next higher rung of the ladder.

After you have filled in the boxes describing the distance determination methods, you should add some reference objects from the Milky Way and the Universe to the distance axis. Add labels for the following objects at the appropriate locations on the distance axis.

- Alpha Centauri
- Center of the Milky Way
- Andromeda Galaxy
- Virgo Supercluster

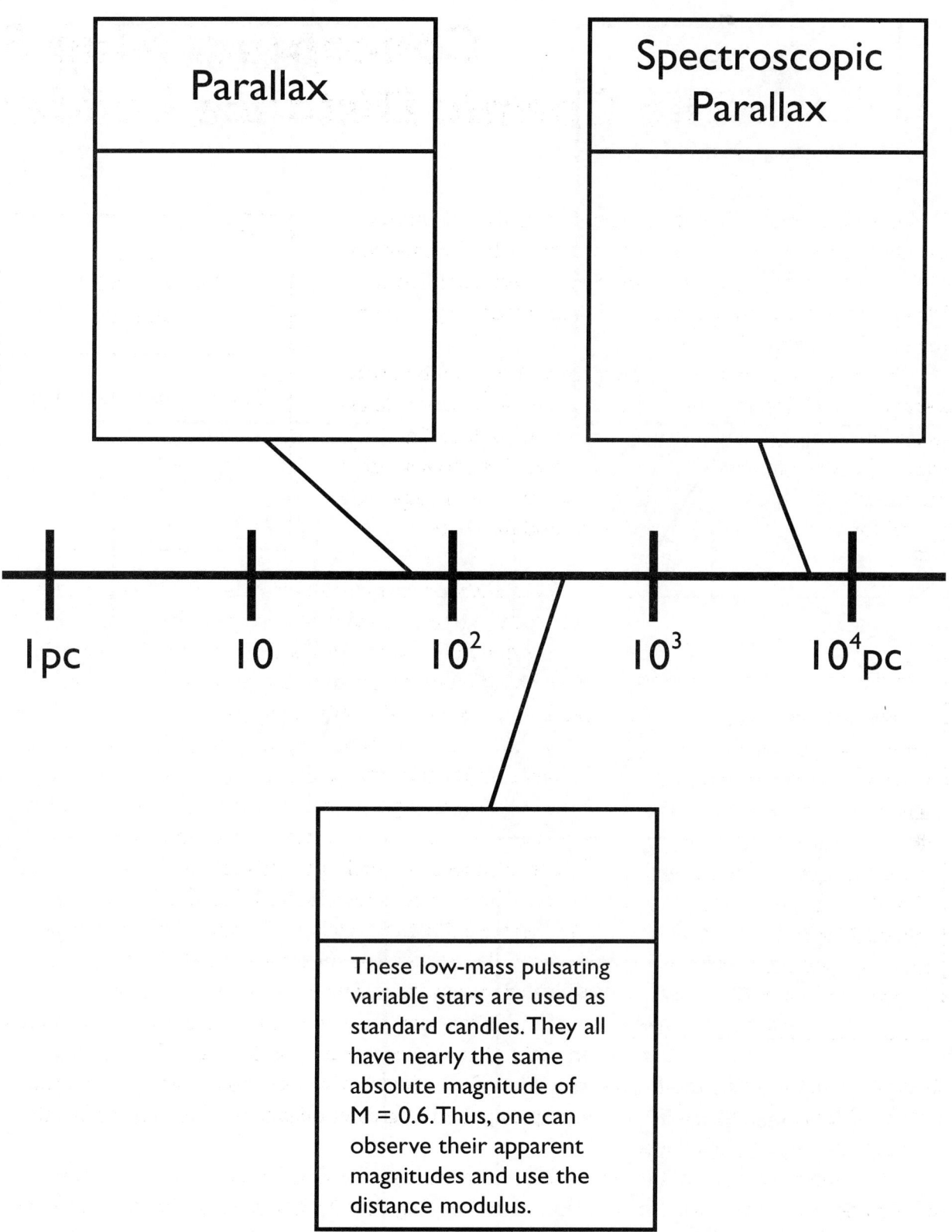

Parallax	Spectroscopic Parallax

1pc   10 10^2 10^3 10^4pc

These low-mass pulsating variable stars are used as standard candles. They all have nearly the same absolute magnitude of M = 0.6. Thus, one can observe their apparent magnitudes and use the distance modulus.

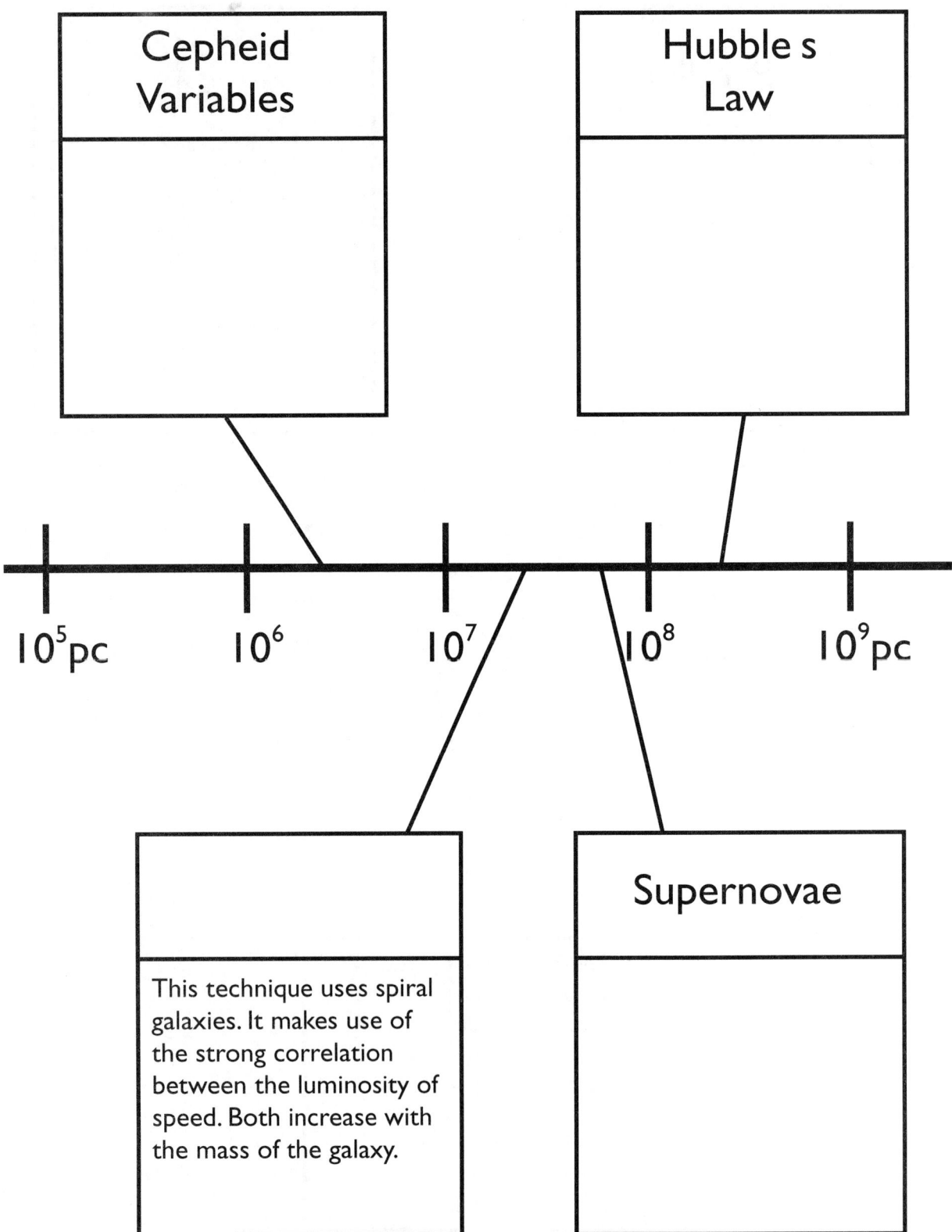

Cepheid Variables

Hubble's Law

10^5pc 10^6 10^7 10^8 10^9pc

This technique uses spiral galaxies. It makes use of the strong correlation between the luminosity of speed. Both increase with the mass of the galaxy.

Supernovae

Unit 25
Active Galaxies and Quasars

Chapter Objectives

Some galaxies have very active, energetic, small nuclei that emit hundreds to thousands of times more energy than an entire normal galaxy emits. The historical development of our knowledge of quasars and related galaxies will be delineated in this chapter. The initial controversy over the distance to and thus the nature of these quasi-stellar objects and the evidence pointing to the cosmological interpretation will be presented. The current consensus concerning the source of the tremendous energy emitted by quasars and active galactic nuclei—supermassive black holes—will be discussed and the supporting evidence explained.

Progress Checklist

1. **Quasars**
❏ Quasars
❏ Redshifts and Distances
❏ Compact Energy Sources
❏ The Energy Problem
❏ Abundance of Quasars
❏ Host Galaxies
2. **Active Galactic Nuclei**
❏ Active Galaxies
❏ Nonthermal Spectra
❏ Radio Galaxies

❏ Seyfert Galaxies
❏ BL Lac Objects
❏ Comparisons with Quasars
3. **A Unified Model of Quasars and AGNs**
❏ Rotating Black Holes
❏ Evidence in M87
❏ Evidence in NGC 4261
❏ Evidence in M84
❏ Evidence in Other Galaxies
❏ Feeding the Black Hole

Keywords

quasars
quasistellar objects, QSOs
cosmological redshift
host galaxy
3C 273
active galaxies
Active Galactic Nuclei, AGNs
radio galaxies

core-halo radio galaxies
lobed radio galaxies
BL-Lacertae objects (blazers)
Seyfert galaxies
optically violent variables (OVVs)
nonthermal emission
synchrotron radiation
relativistic synchrotron jets

polarized light
optical jets
radio jets
supermassive black hole
Virial Theorem
ionization cone
water masers
superluminal motion

Exercise 25-1: Introductory Narrative

Quasars were discovered in the early 1960s. The term *quasar* is a derivative of 1) _____ _____, so called because the first images of quasars were starlike in appearance. However, the spectra of quasars demonstrate extremely large 2) _____ indicating that they have high recessional velocities. By applying 3) _____ one can see that this implies that quasars are very distant objects. Since quasars can be seen at these large distances, they must be very luminous objects. Another interesting characteristic of quasars is that their brightness fluctuates over timescales of several days. Since the fastest possible velocity is the speed of light, quasars can only be several light-days in size. These observations encapsulate the enigma initially posed by quasars: they are about the size of a(n) 4) _____ but produce more energy than a typical 5) _____.

Because almost all quasars are very distant, the 6) _____ times for them must also be very large. This implies that quasars are representative of the Universe of the past and that very few exist today. The distribution of quasars in space is evidence that the Universe 7) _____.

Active galaxies were originally thought to be unrelated to quasars. The 8) _____ of these galaxies exhibit evidence of extremely energetic phenomena such as unusual structure and jets. Their spectra also typically contain 9) _____ radiation at longer wavelengths.

Our understanding of quasars and active galaxies has improved since the 1960s. Today, it is thought that active galaxies and quasars are very similar. Both are thought to be galaxies with very large 10) _____ at their centers and the major difference is the amount of material that is being devoured. Quasars are also more distant, and the stars of the surrounding galaxy are not as easily seen.

Exercise 25-2: True/False Questions

T / F   1. Quasars have very large redshifts, which implies (through Hubble's Law) that they must be at very great distances.

T / F   2. The fact that the luminosity of quasars varies over short timescales implies that they are extremely large objects.

T / F   3. The spectra of quasars exhibit very broad emission lines, further evidence that quasars are traveling away from us at high velocities.

T / F   4. That very few quasars have been found near us in space indicates that very few of them still exist today.

T / F   5. The observation that quasars were far more abundant in the past proves that the Universe evolves over time.

T / F   6. The spectra from active galaxies are dominated by blackbody radiation.

T / F   7. It is likely that the strong radio emissions detected from AGNs are caused by material shot out of compact nuclei by jets.

T / F   8. The major difference between active galaxies and quasars is that quasars are not part of a galaxy.

T / F   9. The existence of large black holes at the centers of active galaxies is supported by evidence for the existence of large amounts of unseen matter at the centers of galaxies.

T / F 10. Accretion disks in the cores of active galaxies can be identified by the redshift of material falling into the black hole.

T / F 11. Quasars are likely to "turn off" if they consume all of the available fuel in their galaxy.

T / F 12. "Turned off" quasars can "turn on" if new matter becomes available to spiral into their accretion disks due to interaction with another galaxy.

Unit 26
Cosmology

Chapter Objectives

Cosmology is the study of the Universe as a whole, as one single object—the ultimate "Big Picture." The Cosmological Principle, which states that on the large scale the Universe is homogeneous and isotropic, is the starting point for this branch of astronomy. Our large scale view of the Universe must be representative if we are to answer the big questions: what is the overall structure of the Universe? what is its history? its fate? In this chapter we will see how the expansion of the Universe (as evidenced by the galactic red shifts) and the application of Einstein's General Theory of Relativity lead naturally to a big bang beginning to our Universe. The possibilities for the ultimate fate of our Universe, whether the expansion will continue forever or rather be stopped and a resulting big crunch will occur, will be explored. The gravitational lensing that is a consequence of Einstein's view of gravity will be described and illustrated and its utility in finding the elusive dark matter that may seal our Universe's fate will be demonstrated.

Progress Checklist

1. **Issues and Implications**
   - ❏ Cosmological Issues
   - ❏ Cosmology & Geometry
   - ❏ Cosmological Principle
   - ❏ Einstein Equations
   - ❏ Cosmological Constant
   - ❏ Fate of the Universe
2. **Gravitational Lensing**
   - ❏ Gravitational Lensing
   - ❏ Lensing of Quasars
   - ❏ Einstein Cross
   - ❏ A Gallery of Lenses
3. **Gamma-Ray Bursts**
   - ❏ Gamma Ray Bursts
   - ❏ Early Observations
   - ❏ Local or Cosmological?
   - ❏ Other Wavelengths
   - ❏ Redshifts & Distances
   - ❏ Models

Keywords

cosmology
expanding universe
big bang
hot big bang
balloon analogy
comoving coordinates
cosmological principle

Friedmann cosmologies
Einstein field equations
cosmological constant
vacuum energy density
cosmic inflation
the big crunch
open universe

closed universe
gravitational lensing
Einstein Cross
gamma ray bursts
BATSE
hypernova

Exercise 26-1: Introductory Narrative

Cosmology is the study of the overall structure and 1) _____ of the Universe. It attempts to answer the big questions like "How did the Universe come to be?" and "What will happen to it in the future?" Thus, cosmology is concerned only with distance scales of 2) _____ of galaxies and timescales of 3) _____ of years.

An observation central to cosmology is that all galaxies are traveling 4) _____ from each other-the Universe is expanding. This suggests that the Universe originated in a high-density, high-temperature state. It then expanded to its present form in an ongoing explosion that is known as the 5) _____. The rate of expansion is slowing over time due to the mutual gravitational interaction of galaxies. Thus, the fate of the Universe (whether the expansion will continue forever) will be determined by the amount of matter in the Universe. The amount of matter necessary to stop the expansion is known as the 6) _____.

The fate of the Universe can also be discussed in terms of the geometry of space. If there is exactly enough matter to halt the expansion, space has no curvature; this is termed the 7) _____ universe. Insufficient matter to stop the expansion implies 8) _____ curvature and an open universe. Sufficient matter to halt the expansion implies positive curvature and a(n) 9) _____ universe. In this scenario the expansion becomes a contraction, and all the galaxies of the Universe would meet in a(n) 10) _____. Although all present methods of calculating the amount of matter in the Universe point to an open universe, the question is still unresolved due to the undetermined nature of 11) _____.

Exercise 26-2: True/False Questions

T / F   1. From any galaxy in the Universe, all other galaxies will be moving away from it with velocities given by Hubble's Law.

T / F   2. Determining the geometry of the Universe requires an understanding of the nature and abundance of dark matter.

T / F   3. If the Universe has positive curvature with enough mass to stop the present expansion, it is said to be open.

T / F   4. The Cosmological Principle states that on the scale of superclusters of galaxies the Universe is homogeneous and isotropic.

T / F   5. If the density parameter is less than one, then the Universe is headed toward a "big crunch."

T / F   6. Recent controversial evidence suggests that the expansion of the Universe is accelerating due to a nonzero vacuum energy density.

T / F   7. The gravitational lensing of quasars is strong evidence that they are really at the distances implied by their redshifts.

T / F   8. Efforts to identify MACHOs (a possible source for the missing baryonic mass) involve the MACHOs eclipsing the stars of the LMC.

T / F   9. Mapping the locations in the sky of gamma-ray bursts suggests that they are occurring in the disk of the Milky Way.

T / F   10. Gamma-ray bursts must release the radiation with very little interaction with matter, or they would be transformed into longer wavelength radiation.

T / F   11. Although there are several theories explaining gamma-ray bursts, all of them involve the collapse of spinning material to form a black hole.

T / F   12. Gamma-ray bursts can release almost as much energy as a nova.

Unit 27
The Early Universe

Chapter Objectives

The Hubble expansion of the Universe and the cosmic microwave background radiation are the two most important pieces of observable data concerning the Universe as a whole. In this chapter we will see how scientists apply these facts and the ideas of elementary particle physics to the first few seconds and minutes of our Universe's history. The few elementary particles (and antiparticles) that could exist in the extremely hot and dense early Universe will be introduced. The important changes that occurred in the early Universe as the temperature and density decreased will be explained. The inflationary versions of the big bang and the additional questions they answer will be discussed. The formation of large scale structure from a relatively smooth early universe in either a top-down or bottom-up scenario will be explored. The necessity for formulating a quantum version of gravity to understand the first small fraction of a second in our Universe's history will be described.

Progress Checklist

1. The Big Bang
- [ ] The Big Bang
- [ ] Cast of Characters
- [ ] Equilibrium & Decoupling
- [ ] The First Three Minutes
- [ ] Subsequent Evolution
- [ ] Triumph of the Big Bang

2. The CBR
- [ ] Discovery
- [ ] Spectrum and Temperature
- [ ] Motion Relative to CMB
- [ ] Isotropy and Anisotropy
- [ ] Fluctuations
- [ ] Constraints on Cosmology

3. Inflationary Universe
- [ ] Problems with Hot Big Bang
- [ ] Unification of the Forces
- [ ] Vacuum Energy

- [ ] Inflationary Expansion
- [ ] Solution of the Problems
- [ ] Fluctuations and Structure

4. Formation of Structure
- [ ] Structure from Uniformity
- [ ] Role of Dark Matter
- [ ] Top-Down Theories
- [ ] Bottom-Up Theories
- [ ] Simulations of Structure Growth
- [ ] Where it Stands

5. The Plank Era
- [ ] The Planck Scale
- [ ] Quantum Gravitation
- [ ] Superstrings and m-Branes
- [ ] Quantum Black Holes
- [ ] Spacetime Foam
- [ ] Breakdown of Current Laws?

Keywords

big bang
radiation dominated universe
matter dominated universe
photon
proton
neutron
electron
positron (antielectron)
neutrino
antineutrino
antimatter
quarks
gluons
thermal equilibrium
decoupled
freezout
quark confinement
confinement transition
deuterium bottleneck
baryon
recombination transition

steady state model
perfect cosmological
    principle
cosmic background radiation
    (CBR)
blackbody curve
isotropy
anisotropy
horizon problem
dipole anisotropy
inflationary universe
flatness problem
magnetic monopole problem
Grand Unified Theories
    (GUTs)
elementary particle physics
lightcone
particle horizon
event horizon
strong interaction
electromagnetic force

weak force
gravitational force
Superunified Theories
Standard Model
Standard ElectroWeak
    Theory
spontaneous symmetry
    breaking
Planck scale
inflationary epoch
phase transition
vacuum energy
cosmological constant
top-down theories
bottom-up theories
Planck era
quantum gravitation
superstring theory
spacetime foam

Exercise 27-1: Introductory Narrative

The Universe began in a state of very high temperature and density in a gigantic 1) _____
_____ known as the big bang. Most of the energy of the early Universe
was in the form of 2) _____. The simplicity of the early Universe can be
seen through the small number of fundamental particles that existed: photons, protons, neutrons, electrons, neutrinos, and various 3) _____. Radiation and matter
were in thermal equilibrium, and energy moved freely between the two forms in events
of pair creation and pair 4) _____. As the temperature fell, the reaction
rates for certain types of particles also fell, and eventually one by one the particles became
5) _____ from the general thermal equilibrium. It was still so hot that a
typical photon had sufficient energy to break apart a neutron that had fused with a proton. This
barrier to nuclear fusion is known as the deuterium 6) _____. Since free
neutrons are not stable, their numbers were decreasing at this time. Once the temperature fell
enough for deuterium to form, fusion of nuclei up to helium-4 occurred rapidly. This pre-
served the relative abundance of protons and neutrons at that time. Nuclear fusion was unable
to build nuclei larger than helium-4 because there were no stable 7) _____
or mass-8 isotope nuclei. After 1000 years, when the universe had cooled to a temperature of
100,000 K, most of the energy was in the form of 8) _____.

The next major event in the big bang occurred at a time of 300,000 years when the Universe had cooled to a temperature of about 3000 K. It was now cool enough for electrons to remain bound to nuclei. Light no longer interacted with free electrons, and the Universe, which had been opaque, now became 9) _____. We still detect the radiation released at this time, which is called the 10) _____. The discovery of this radiation offered definitive proof to modern cosmologists that the big bang theory was correct. We are part of the big bang, an event that is still going on today as the Universe continues to expand and cool.

Exercise 27-2: True/False Questions

T / F   1. The very early Universe was *matter dominated* because the energy was primarily in the form of nonrelativistic matter.

T / F   2. When a particle meets its antiparticle, they annihilate each other, meaning that they completely disappear with no trace.

T / F   3. The term *observable universe* is just another way of saying the entire universe.

T / F   4. An imbalance between the number of protons and neutrons occurred in the early Universe because free neutrons are less stable than protons and it was too hot for neutrons to form deuterium.

T / F   5. The nuclei of heavy elements such as carbon and nitrogen formed in the early Universe in a manner very similar to how they are formed in the cores of massive stars.

T / F   6. The relative abundances of deuterium and helium in the oldest stars serve as a good indicator of conditions in the early Universe.

T / F   7. The big bang theory predicts that there should be considerably more helium in the Universe than astronomers observe.

T / F   8. Since the time of the release of the microwave background radiation, its wavelength has increased and the number density of its constituent photons has decreased.

T / F   9. The "dipole anisotropy" of the COBE data is due to the Earth's orbital motion about the Sun.

T / F   10. It is difficult to explain how galaxies could have grown from the very small fluctuations in the CBR without involving more complex structure in dark matter.

T / F   11. The difficulty in explaining how parts of the CBR from opposite directions in the sky, which could never have been in contact, are at exactly the same temperature is called the horizon problem.

T / F   12. If a magnet is cut in half, north and south magnetic monopoles will be produced.

T / F   13. The top-down theory of galaxy formation predicts that masses the size of star clusters form first and later build galaxies through collisions and mergers.

T / F   14. The present observational evidence concerning the formation of structure favors a bottom-up assembly and an important role for cold dark matter.

T / F   15. Superstrings (which are theorized to be the fundamental building blocks of the Universe) are one-dimensional objects.

Conceptual Map 6
Look-Back Time

In this assignment we will again work with a logarithmic distance axis as we did in Conceptual Map 5. However, this time our axis will be labeled in Light Years (Ly) from 1 million light years to over 10 billion light years. Thus our distance axis begins at the Milky Way and extends outward to greater distances. Our goal is to note the differences in the types of objects we see as we look to greater and greater distances.

The general format of this Conceptual Map is a box with a small area on top to contain the name of the typical object seen at a certain distance and a larger area below to describe the characteristics of the object. In each of these boxes the name of the object is already completed for you and you are expected to add the description.

Typical Object Seen at a Certain Distance
Description of Object

A line is drawn connecting the box to an appropriate distance on the axis where this type of object is typically found. One should note that this is just a typical distance. Normal galaxies, active galaxies, and quasars are all found at a wide range of distances which overlap each other. *You should now complete the description boxes connected to the distance axis.*

It is important to realize that the axis represents not only distance, but also look-back time. When you look at objects found at great distances, it has taken light a very long time to travel this far. You are effectively looking back in time. This was the motivation behind labeling the axis in light years. A distance in light years reflects a look-back time in years. Thus, the fact that we see very different objects at different distances/look-back times illustrates that the universe is evolving over time. *Add the terms **Big Bang** and **Formation of Galaxies** to the appropriate location on the axis of your map.*

One can carry the meaning of the distance axis one step further in hopes of understanding the differences in the types of objects. We know from Hubble's Law that the universe is expanding. So, when we look to large distances we are looking back in time when the universe was smaller and the density of the material in the universe was larger. (The actual look-back time depends on both the Hubble Constant and the deceleration parameter. You should experiment with the Redshift Calculator **(IC 24-9)** to gain some familiarity with these relationships.) Thus, (in general) galaxies appear different in the past because there was more material around for the black holes in the cores of active galaxies to devour and considerably more material for quasars at greater distances. The average distances between galaxies would also be smaller and they would interact considerably more often than they do at present which can also cause "feeding" of the black holes.

Your Conceptual Map is also useful for thinking about the homogeneity and isotropy of the universe. The sequence depicted as one looks at greater distances-Normal Galaxies, Active Galaxies, Quasars, and Microwave Background Radiation-would be exactly the same for any observer located anywhere in the universe regardless of the direction in which they were looking. Our universe is truly homogeneous and isotropic in that every observer in the universe sees themselves at the center of their own observable universe. This is one of the reasons cosmologists state that "the universe has no center and no edge."

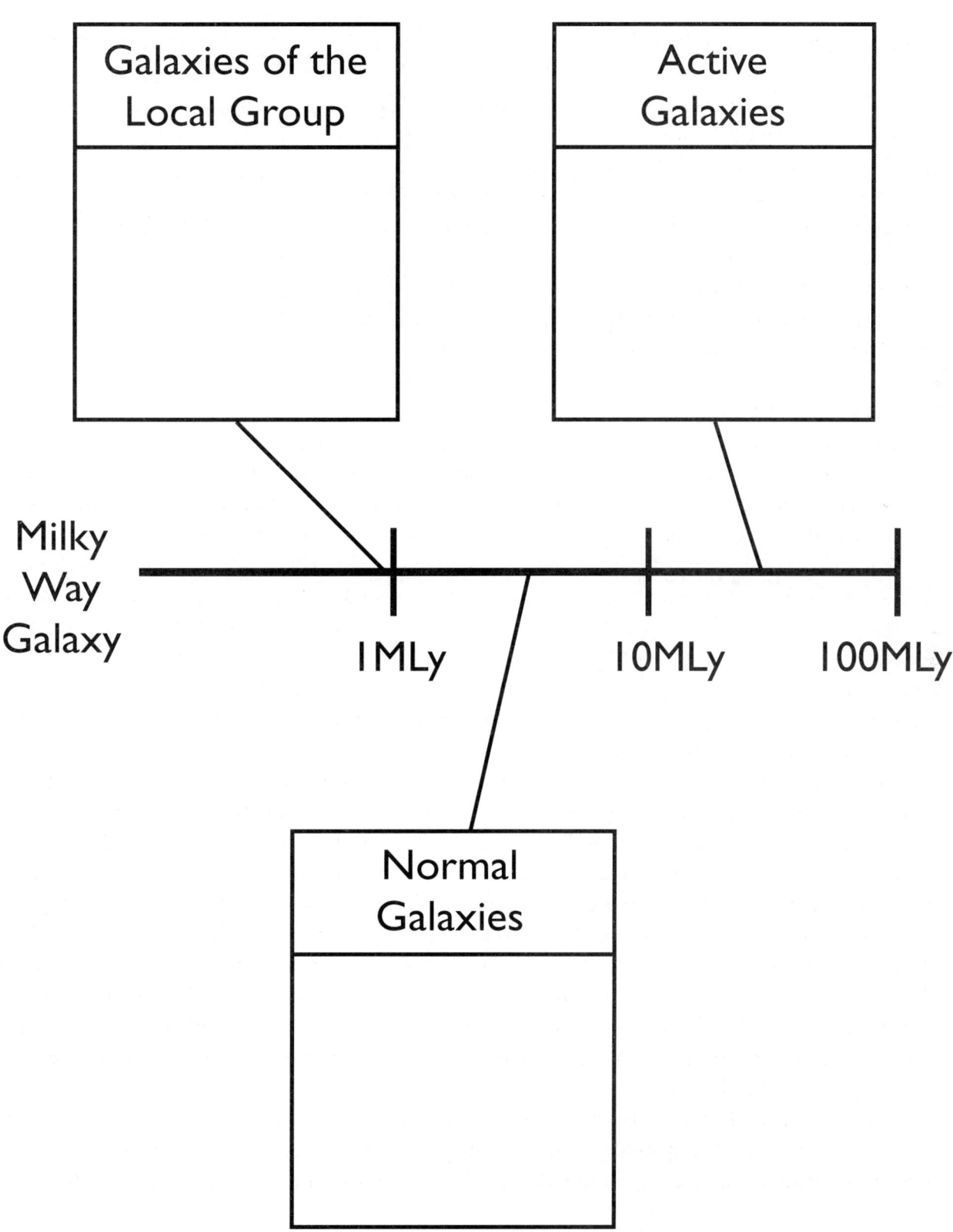

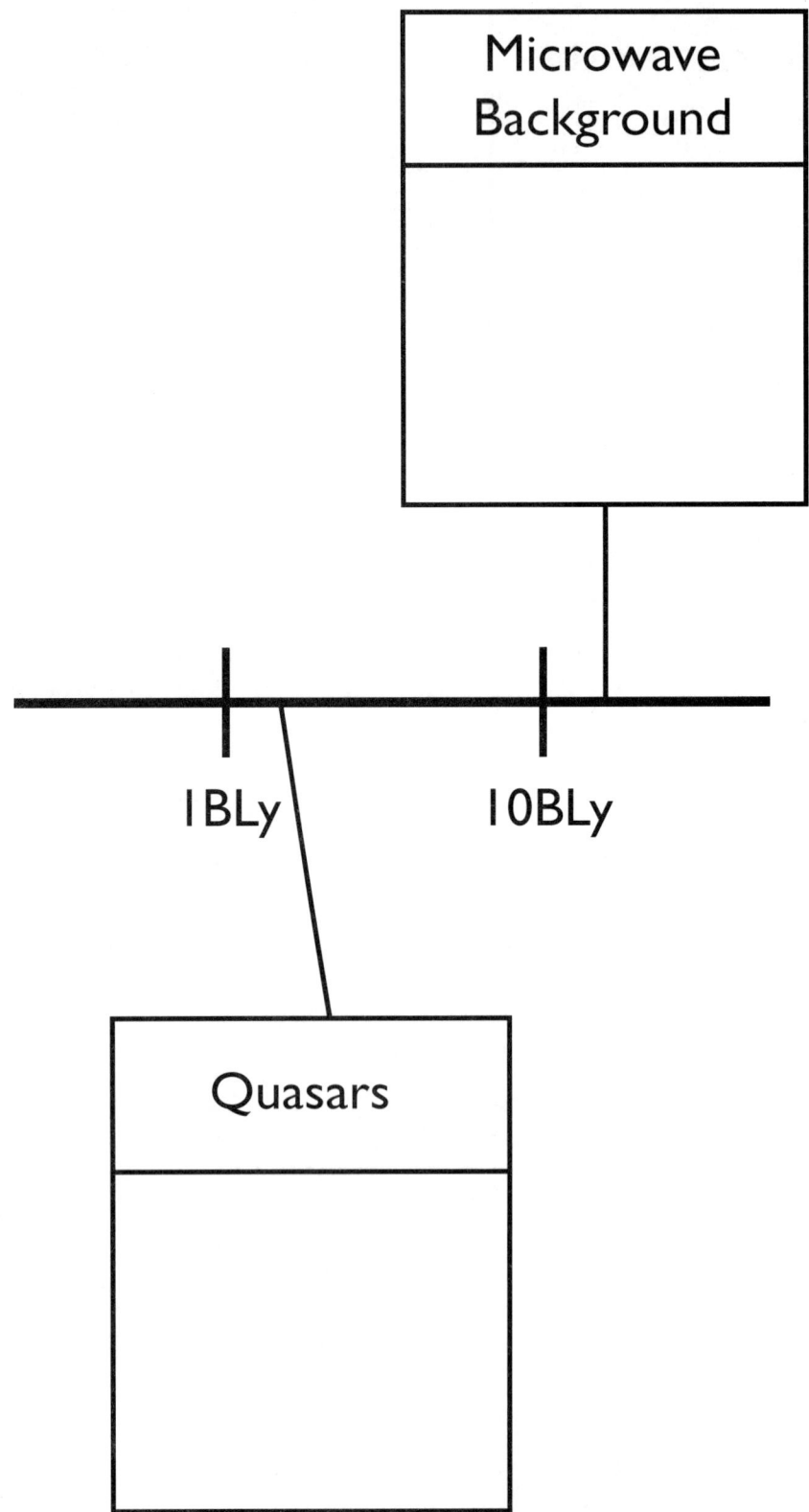

Unit 28
Life in the Universe

Chapter Objectives

It is a fascinating question: Are we alone? How do scientists consider the topic of life elsewhere in the Universe without dealing entirely in pure speculation or science fiction? The basic principles underlying the search for extraterrestrial life will be outlined in this chapter. The definition of life and its most fundamental biological components will be discussed. The probability of other solar systems with planets within their habitable zone (ecosphere) will be investigated. The possibility of communicating with an extraterrestrial intelligent civilization will be explored and the criteria used to decide the proper method of communication will be discussed.

Progress Checklist

1. **Biological Conditions**
❑ Defining Life
❑ The DNA Code
❑ Evolution of Life
❑ Origin of Life on Earth
❑ Timescales
❑ Is Carbon-Based Life Unique?
2. **Astronomical Conditions**
❑ Solar Systems
❑ Habitable Zones
❑ Timescales
❑ How Many Civilizations?
3. **Communicating to Extraterrestrial Civilizations**
❑ Distance and Time
❑ UFOs?
❑ Sending Messages
❑ Radio Searches
❑ SETI
❑ What if They Reply?

Keywords

life
deoxyribonucleic acid (DNA)
cell
adenine (A)
cytosine (C)
guanine (G)
thymine (T)

genes
chromosomes
gene mutation
ribonucleic acid (RNA)
evolution
natural selection
carbon-based chemistry

habitable zone (ecosphere)
Drake equation
UFOs
water hole
SETI
Project Phoenix

Exercise 28-1: Introductory Narrative

To accurately assess the likelihood of life existing on other planets, we must first make sure we understand the nature of life on Earth. Life began in the Earth's oceans about 1) _____ _____ years ago. Life is based upon the chemistry of the 2) _____ _____ atom. The building blocks of life (amino acids, etc.) were formed from the interaction of the Earth's primitive atmosphere, UV radiation, and lightning. The building blocks formed larger organic molecules through a process known as 3) _____ _____. At some point a molecule with the ability to reproduce itself was produced; that may be considered the first life. All living things carry information about their heritable traits in a long twisting molecule known as 4) _____. In successive generations of the species, these data can change through a process known as 5) _____. Some of these changes or 6) _____ are advantageous to the species due to a changing environment and help the species to survive in a process known as 7) _____.

Do the conditions that were necessary for the development of life on Earth exist on planets of other solar systems? Liquid water was certainly essential. Astronomers define the region around a star where liquid water can exist as the 8) _____, the size of which increases substantially for hotter stars. However, these hot stars live only a short time on the main sequence. Taking both of these factors into account suggests that spectral type 9) _____ stars are the most likely to have life evolve on planets orbiting them. One can attempt to estimate the number of intelligent civilizations in our galaxy using the 10) _____ equation.

Even if other intelligent civilizations have evolved in our galaxy, they are probably not near us in either space or time. Thus, sending messages as was done with the 11) _____ spacecraft in 1977 is one possible means of opening communication. A method more likely to be successful is sending and receiving radio signals. A signal was sent from the Arecibo Radio Observatory toward the globular cluster M13 in 1974. The NASA project SETI scanned the skies for signals, and its successor project 12) _____ still does so today.

Exercise 28-2: Using the Drake Equation

The Drake equation is an attempt to calculate how many technically advanced civilizations might exist in our galaxy (or any other) with which we potentially could communicate. The number of civilizations is given by the product of the factors in the table to the right. The Drake equation is really more a model for thinking about the problem than a realistic method of making this calculation. Several of the factors in the Drake equation are extremely uncertain, and there is considerable debate among astronomers concerning their values.

$$N = N_S \cdot F_P \cdot N_H \cdot F_L \cdot F_I \cdot F_S$$

N is the number of communicative civilizations in the galaxy.

N_S is the number of stars in the galaxy.

F_P is the fraction of those stars with planets.

N_H is the average number of those planets in the habitable zone.

F_L is the fraction of those that evolve life.

F_I is the fraction of those for which life evolves to intelligence

F_S is the fraction of the star's life in which the civilization is communicative.

The Drake Equation

The table below contains some consensus optimistic and pessimistic values for all of the parameters of the Drake equation and a column for you to enter your estimates. You should enter your estimates for each of the parameters in column 4, and state in column 5 whether you are closer to the optimists' or pessimists' estimate for this value; then multiply your column 4 values together to estimate the number of technically advanced civilizations in our galaxy. Your result will be just as valid as any astronomer's estimate.

Variable	Pessimistic Estimates	Optimistic Estimates	Your Estimates	Closer to Optimistic or Pessimistic
N_S	2×10^{11}	2×10^{11}		
F_P	0.01	0.5		
N_H	0.01	1		
F_L	0.01	1		
F_I	0.01	1		
F_S	10^{-8}	10^{-4}		
N	2×10^{-5}	1×10^{7}		

Are you an optimist or a pessimist? Which parameters do you think have the most uncertainty associated with them?

Exercise 28-3: True/False Questions

T / F   1. Life on the planet Earth is based on the complex chemistry of the element carbon.

T / F   2. Genes are combinations of chromosomes, which along with many proteins make up a single DNA molecule.

T / F   3. Damage from cosmic rays, solar UV, or natural radioactivity and alteration of DNA by chemicals in the environment all are possible sources of genetic mutation.

T / F   4. Laboratory experiments simulating the interaction of the Earth's primitive atmosphere and an energy source produce many of the chemical building blocks of life.

T / F   5. BillyBuzz the dragonfly might evolve larger wings in order to fly faster to escape the predatory birds that live near his pond.

T / F   6. Life is just as likely to develop on a planet in the habitable zone around a B2 spectral type star as around any other star.

T / F   7. Life is just as likely to develop on a planet orbiting a binary star system as on a planet orbiting any other star.

T / F   8. Life is just as likely to develop on a planet orbiting a star in a highly elliptical orbit as in any other scenario.

T / F   9. It is possible that Jupiter's moon Europa may have subsurface liquid oceans because Europa is within the Sun's habitable zone.

T / F   10. Spaceflight to other star systems is unlikely in the near future due to the extreme distances involved and the limitation that velocities must be less than the speed of light.

T / F   11. Radio waves with wavelengths between 18 and 21 cm are logical ways to send signals to extraterrestrials.

Appendix
Solutions to Exercises

Exercise 17-1: Introductory Narrative

1. helium   2. core   3. convection   4. hydrostatic equilibrium   5. luminosity
6. photosphere   7. granulation   8. chromosphere   9. corona   10. magnetic
11. Sunspots   12. flares   13. solar wind

Exercise 17-2: True/False Questions

1. **T**   2. **T**   3. **F**, The solar constant is only the amount of energy passing through 1 square meter of the surface of that sphere.   4. **F**, Radiative transport is the most important mechanism in the inner parts of the Sun. Convection only becomes significant near the Sun's surface.   5. **T**   6. **F**, The edge of the Sun will be darker, a phenomenon known as limb darkening.   7. **T**   8. **T**   9. **F**, The number of sunspots varies with an 11-year periodicity.   10. **T**   11. **F**, The Babcock model states that sunspots occur due to convection being inhibited.   12. **T**   13. **T**

Exercise 18-1: Introductory Narrative (Stellar Energy Production)

1. energy   2. nuclear (or fusion)   3. stable   4. positively   5. Coulomb barrier
6. temperature   7. helium   8. proton-proton   9. convection

Exercise 18-2: Introductory Narrative (Stellar Parameters)

1. parallax   2. Hipparchos   3. proper motion   4. Doppler shift   5. brightness
6. distance   7. absolute magnitude   8. temperature   9. B   10. K   11. color index
12. Hertzsprung-Russell   13. radius

Exercise 18-3: Using Parallax

3. The length is approximately 33.5 blocks.   4. $(33.5 \ blocks)\left(\dfrac{20 \ m}{1 \ block}\right) = 670 \ m$

Exercise 18-4: Using Magnitudes

1. α (smallest m)   2. κ (largest m)   3. ι $(16 \Rightarrow m_2 - m_1 = 3)$ 4. β or γ $(100 \Rightarrow m_2 - m_1 = 5)$ 5. $16(m_2 - m_1 = 6 - 3 = 3)$ 6. $40(m_2 - m_1 = 5 - 1 = 4)$ 7. 4 (at 10 pc, $m = M$)
8. $-2(m - M = 3 \Rightarrow 1 - M = 3)$ 9. $160 \ [m - M = 2 - (-4) = 6]$ 10. $16 \ [m - M = 1 - 0 = 1]$ 11. $-1.5 (140 \ \text{pc} \Rightarrow m - M = 5.5)$ 12. $9[m - M = 2 - (-7) = 9]$

Exercise 18-5: The HR Diagram

1. Spica   2. Betelgeuse   3. Rigel   4. Procyon B   5. Betelgeuse-Large stars are found to the upper right of the HR Diagram.   6. Sirius B or Procyon B (It should be noted that some astronomers don't refer to white dwarfs as stars because they no longer produce energy through nuclear fusion reactions. They are referred to as compact objects-objects that used to be stars.)   7. Spica, Altair, Procyon A, and the Sun (Vega and Sirius are very close) 8. Vega, Sirius A, or Deneb   9. Rigel, Deneb, or Betelgeuse   10. Betelgeuse-A star must be very cool to have molecular lines in its spectrum.

Exercise 18-6: Spectroscopic Parallax

Star 1: Spectral Type = B5
Luminosity Class = V
Absolute Magnitude = -2
Distance Modulus = 11
Distance (in pc) = 1600
Star 2: Spectral Type = B5
Luminosity Class = Ia or Ib
Absolute Magnitude = -7
Distance Modulus = 11
Distance (in pc) = 1600
Star 3: Spectral Type = K2
Luminosity Class = II
Absolute Magnitude = -2
Distance Modulus = 8
Distance (in pc) = 400

Star 4: Spectral Type = O8
Luminosity Class = V
Absolute Magnitude = -5
Distance Modulus = 12
Distance (in pc) = 2500
Star 5: Spectral Type = K2
Luminosity Class = III
Absolute Magnitude = 0
Distance Modulus = 10
Distance (in pc) = 1000
Star 6: Spectral Type = G0
Luminosity Class = IV
Absolute Magnitude = 2
Distance Modulus = 6
Distance (in pc) = 160

Exercise 18-7: True/False Questions

1. **F**, The energy from gravitational contraction is very small.   2. **F**, A fusion reaction involves the combining of two small nuclei.   3. **T**   4. **T**   5. **F**, The CNO Cycle is important in massive stars.   6. **T**   7. **T**   8. **F**, Neutrinos interact very weakly with matter and often escape from stars without interacting with matter at all.   9. **F**, The space-based Hipparchos satellite measured parallax to an accuracy of a thousandth of an arc second. 10. **T**   11. **F**, The V filter is very similar to the eye, while the B filter is most similar to photographic film.   12. **T**   13. **F**, Different spectral types are predominantly due to stars having different surface temperatures. Composition is a minor influence.   14. **T**

Exercise 19-1: Introductory Narrative

1. visual   2. astrometric   3. spectroscopic   4. radial velocity curves   5. eclipsing
6. 90°   7. light curve   8. radii   9. accretion disk   10. Open   11. Globular
12. Population II

Exercise 19-2: Eclipsing Binary Light Curves

1. To simplify matters, we will make an assumption concerning the orbital velocities of the stars. Assume that star A is moving out of the page toward the observer in frame I. Answers are given in the first orbital cycle.

    Frame I—5 days

    Frame II—This could be either 10 days or 33 days.

    Frame III—14 days

    Frame IV—16 days

    Frame V—This could be either 20 days or 25 days.

    Frame VI—29 days

2. Approximately 33 days
3. Primary eclipse—1.0 magnitude; secondary eclipse—0.4 magnitude

Exercise 19-3: Spectroscopic Binary Stars

1. Frame I—5 days

    Frame II—20 days

    Frame III—13 days

    Frame IV—0 days

2. Approximately 26 days
3. −15 km/s
4. Star B is the more massive because its radial velocity has a smaller range of values.

Exercise 19-4: The Ages of Open Clusters

1. Upper Left Cluster—Turnoff point is approximately 9750 K.
   —Age is approximately 400 My.
   —This cluster contains two blue stragglers.
2. Upper Right Cluster—Turnoff point is approximately 14,000 K.
   —Age is approximately 100 My.
3. Lower Left Cluster—Turnoff point is approximately 16,000 K.
   —Age is approximately 75 My.
4. Lower Left Cluster—Turnoff point is approximately 6,500 K.
   —Age is approximately 6,000 My.
   —This cluster contains two blue stragglers.

Exercise 19-5: True/False Questions

1. **F**, Both stars orbit around the center of mass. Although Sirius A will be closer to this point than Sirius B, it certainly is not stationary.   2. **T**   3. **F**, This relation is only valid for main sequence stars.   4. **F,** One can only determine the total mass of the pair of stars. 5. **T**   6. **F**, When the separation between source and observer is increasing, spectral lines will be redshifted.   7. **F**, For eclipses to occur, the angle of inclination i must be near 90°. 8. **F**, The B8 star has a much higher surface temperature than the K2 star; thus, the deeper eclipse must occur when the K2 star eclipses it.   9. **T**   10. **F**, Novae involve the accretion of material onto the surface of a white dwarf.   11. **T**   12. **T**   13. **T**   14. **F**, Metals are formed inside stars over time. The stars in globular clusters are very old but contain very small amounts of metals because not many metals had been made yet at the time the stars formed.   15. **T**

Exercise 20-1: Introductory Narrative

1. molecular clouds   2. Jeans   3. shock waves   4. fragmentation   5. energy
6. energy transport   7. cocoon   8. bipolar flows   9. 30 million   10. evolutionary track   11. hydrogen   12. 0.8

Exercise 20-2: Luminosities and Lifetimes of Main Sequence Stars

Star #1: Proxima Centauri $M = 0.1 M_\odot$

$$L = (M)^{3.5} = (0.1 M_\odot)^{3.5} = 3.2 \times 10^{-4} L_\odot$$

$$T = \frac{1}{M^{2.5}} = \frac{1}{(0.1 M_\odot)^{2.5}} = 316 T_\odot$$

$$316 T_\odot = 316 T_\odot \left(\frac{1 \times 10^{10} \ years}{1 T_\odot} \right) = 3.12 \times 10^{12} \ years$$

Star #2: Rigel $M = 10 M_\odot$

$$L = (M)^{3.5} = (10 M_\odot)^{3.5} = 3162 L_\odot$$

$$T = \frac{1}{M^{2.5}} = \frac{1}{(10 M_\odot)^{2.5}} = 316 \times 10^{-3} T_\odot$$

$$3.16 \times 10^{-3} T_\odot = 3.16 \times 10^{-3} T_\odot \left(\frac{1 \times 10^{10} \ years}{1 T_\odot} \right) = 3.16 \times 10^{7} \ years$$

Exercise 20-3: True/False Questions

1. **T**   2. **F**, The temperature will rise.   3. **F**, Hydrostatic equilibrium refers to a state of balance between expansion due to pressure and contraction due to gravity. A star is not in hydrostatic equilibrium if it is expanding or contracting.   4. **T**   5. **T**   6. **T**   7. **F**, UV radiation merely evaporates the low-density material surrounding the EGGs so that they may be seen.   8. **T**   9. **F**, Collapsing protostars are fully convective. Energy cannot flow efficiently by radiative transport.   10. **F**, Massive stars will collapse much more quickly than low-mass stars.   11. **T**   12. **F**, The width is due to differing compositions, which affect opacities.   13. **T**   14. **F**, Although lithium is useful, it is because it is destroyed in stellar fusion reactions. Thus, if lithium is present, it is assumed that the object has never had nuclear reactions.   15. **T**

Exercise 21-1: Introductory Narrative

1. 10 billion   2. upper right   3. hydrogen   4. red giant   5. helium flash
6. horizontal branch   7. red giant   8. planetary nebula   9. left   10. electron pressure
11. white dwarf   12. Earth

Exercise 21-2: Using Pulsating Variable Stars as Distance Indicators

Star #1: $m - M = 9.5 - 0.5 = 9$
   This corresponds to a distance of 630 pc.
Star #2: From the graph one can estimate that $M = -1$.
   $m - M = 4 - (-1) = 5$
   This corresponds to a distance of 100 pc.
Star #3: From the graph one can estimate that $M = -4$.
   $m - M = 6 - (-4) = 10$
   This corresponds to a distance of 1000 pc.

Exercise 21 -3: True/False Questions

1. **T**   2. **T**   3. **F**, The Sun will first burn hydrogen in a shell surrounding the helium ash core. It is necessary for the helium core to collapse and get much hotter before helium fusion can begin.   4. **T**   5. **T**   6. **F**, Planetary nebulae really have nothing to do with planets. 7. **F**, There are no fusion reactions occurring in a white dwarf.   8. **F**, The 1.4-1.5 solar mass value is an upper limit for the mass of a white dwarf.   9. **T**   10. **F**, Although Cepheids are useful as distance indicators because of their period-luminosity relation, the luminosity gets larger for stars with larger pulsation periods.   11. **F**, Type I supernovae involve accretion onto a white dwarf in a binary system. There are no hydrogen lines because there is very little hydrogen in a white dwarf.   12. **T**   13. **T**   14. **F**, The most stable nuclei are those in the "iron peak" with mass numbers between 55 and 60.   15. **F**, The r-process requires an abundance of heavy nuclei like iron and a strong neutron source. These conditions are likely to be present in a Type II supernova.

Exercise 22-1: Introductory Narrative

1. white dwarfs   2. 1.44 solar masses   3. protons   4. a city   5. magnetic
6. rapidly   7. synchrotron   8. pulsar   9. lighthouse   10. singularity   11. event
horizon   12. charge   13. X-rays

Exercise 22-2: True/False Questions

1. **T**   2. **F**, The lower limit for the mass of a neutron star is 1.44 solar masses. Stellar remnants with lower masses than that can become white dwarfs.   3. **F**, They are likely the imploded cores of Type II supernovae, which are formed from massive stars. Type I supernovae involve white dwarfs in binary systems.   4. **T**   5. **T**   6. **F**, A nova occurs when matter accretes onto a white dwarf in a binary system. Although the phenomenon involving a neutron star is very similar, it is known as an X-ray burster.   7. **T**   8. **F**, The two axes generally are not aligned.   9. **F**, The rate of spinning slows as a pulsar get older because it is slowly radiating away energy.   10. **T**   11. **F**, Millisecond pulsars spin so rapidly that they must have gained angular momentum from some external source-probably a binary companion.   12. **T**   13. **T**   14. **F**, The statement refers to the fact that all information about the matter that makes up a black hole is lost. Black holes have very few distinguishing characteristics.   15. **T**

Exercise 23-1: Introductory Narrative

1. 200,000,000,000   2. Harlow Shapley   3. Edwin Hubble   4. central bulge
5. slowly   6. low   7. II   8. rapidly   9. I   10. rotation curve   11. dark matter

Exercise 23-2: The Components of the Milky Way Galaxy

1. D   2. H   3. D   4. H   5. D   6. D   7. H   8. H   9. H   10. D   11. H
12. H   13. D   14. D   15. H

Exercise 23-3: True/False Questions

1. **T**   2. **F**, Galactic latitude is being described.   3. **F**, All stars produce more visual light than infrared. The reason we can see the central region of the Milky Way better in infrared is that it isn't obscured as much by the intervening dust.   4. **F**, A population I star is high in metals that are produced by nucleosynthesis over time. Thus, a population I star is likely to be much younger.   5. **T**   6. **F**, Disk stars have circular coplanar orbits. Halo stars have randomly oriented elliptical orbits.   7. **T**   8. **F**, The spiral arms are regions of active star formation, and they appear bright because of the hot massive stars that form there.   9. **T**   10. **T**   11. **F**, Rotation curves definitely suggest some type of dark yet gravitational matter, but there are many other possible explanations for this than black holes.   12. **T**   13. **F**, HII regions have hydrogen atoms that have been ionized. HII regions are much hotter than clouds with molecular hydrogen.   14. **F**, Interstellar reddening is due to the dust grains of the interstellar medium, which are much larger than gas molecules.   15. **T**

Exercise 24-1: Introductory Narrative

1. irregular   2. bars   3. spherical (round)   4. large   5. loosely   6. Hubble Tuning Fork Diagram   7. dust   8. star formation   9. evolve   10. superclusters   11. gravitational   12. dark matter

Exercise 24-2: Distances to Galaxies

Galaxy #1: $m - M = 20.5 - (-17.5) = 38 \rightarrow d = 4 \times 10^8$ pc = 400 Mpc

Galaxy #2: $m - M = 14.3 - (-19.7) = 34 \rightarrow d = 6.3 \times 10^7$ pc = 63 Mpc

Exercise 24-4: True/False Questions

1. **F**, The Hubble Tuning Fork Diagram merely classifies galaxies on the basis of structure. It says nothing about galaxy evolution.   2. **T**   3. **F**, Spiral galaxies have far more gas and dust than do elliptical galaxies.   4. **F**, Spiral arms are visible because they are regions of active star formation and have massive, extremely luminous stars in them. These star form there and don't live very long. Thus, they are seen only in the spiral arms because they die before they can move out.   5. **F**, There is a substantial observational bias in favor of spiral galaxies because they are on average much brighter than elliptical galaxies. Thus, we see many more spiral galaxies even though elliptical galaxies are considerably more abundant.   6. **T**   7. **T**   8. **F**, A mass to light ratio of 400 implies the presence of considerable material that is not producing light-dark matter.   9. **T**   10. **F**, The motion toward the Great Attractor is a small local (peculiar) effect superimposed on the expansion of the Universe known as the Hubble flow.   11. **T**   12. **T**   13. **T**   14. **F**, A typical mature spiral galaxy like the Milky Way produces 2 or 3 new stars per year. In a starburst galaxy the rate is about 100 times larger.

Exercise 25-1: Introductory Narrative

1. quasi-stellar   2. redshifts   3. Hubble's Law   4. solar system   5. galaxy   6. look-back   7. evolves   8. cores (centers)   9. nonthermal   10. black holes

Exercise 25-2: True/False Questions

I. **T**   2. **F**, For a quasar's luminosity to change over a short period of time, it must be a small object because whatever physical process is causing the change in luminosity must travel slower than the speed of light.   3. **F**, Broad lines are caused by random velocities.   4. **T**   5. **T**   6. **F**, Most of the radiation coming from active galaxies is nonthermal synchrotron radiation.   7. **T**   8. **F**, Quasars are thought to be the superluminous centers of distant galaxies; they are so far away that the surrounding galaxy cannot be seen as it can for nearby active galaxies.   9. **T**   10. **F**, The radial velocity near the cores of active galaxies shows an "S pattern" (both a redshift and a blueshift) from the material rotating in the accretion disk.   11. **T**   12. **T**

Exercise 26-1: Introductory Narrative

1. origin   2. superclusters   3. billions   4. away   5. big bang   6. critical density
7. flat   8. negative   9. open   10. bigcmnch   11. dark matter

Exercise 26-2: True/False Questions

1. **T**, (Ignoring any peculiar motions of cluster members)   2. **T**   3. **F**, What is described is a closed universe.   4. **T**   5. **F**, For the Universe to collapse back on itself, the density parameter must be greater than one.   6. **T**   7. **T**   8. **F**, MACHOs could be detected by the lensing (short term brightening) of the light from distant stars.   9. **F**, The distribution of gamma-ray bursts is isotropic, and redshift observations indicate that they are not limited to our galaxy.   10. **T**   11. **T**   12. **F**, Gamma-ray bursts release as much or more energy as a supernova.

Exercise 27-1: Introductory Narrative

1. explosion   2. radiation   3. antiparticles   4. annihilation   5. decoupled
6. bottleneck   7. mass-5   8. matter   9. transparent   10. cosmic background radiation

Exercise 27-2: True/False Questions

1. **F**, The early Universe was radiation dominated.   2. **F**, The annihilation must still satisfy the conservation of mass-energy; thus, photons with equal energy are produced.   3. **F**, The observable universe (the part we can observe) is a subset of the entire universe.   4. **T**
5. **F**, Nuclei more massive than helium-4 could not form in the early Universe in any appreciable amount.   6. **T**   7. **F**, There is good agreement between estimates of helium abundance from cosmology and those from observation.   8. **T**   9. **F**, The dipole anisotropy is due to the motion of the Local Group of galaxies, which move at about 600 km/s relative to the CBR.
10. **T**   11. **T**   12. **F**, One simply gets two smaller magnets. It is not possible to form a magnetic monopole today.   13. **F**, What is described is the bottom-up theory.   14. **T**
15. **T**

Exercise 28-1: Introductory Narrative

1. 3.0-3.6 billion   2. carbon   3. chemical evolution   4. DNA   5. evolution
6. mutations   7. natural selection   8. habitable zone   9. F-G-K   10. Drake
11. Voyager   12. Phoenix

Exercise 28-3: True/False Questions

1. **T**   2. **F**, Actually, chromosomes are combinations of genes that make up DNA.   3. **T**
4. **T**   5. **F**, Evolution is a very time-consuming process. BillyBuzz cannot evolve in any way. However, if he happens to have larger wings than other dragonflies, he is more likely to escape predatory birds and pass that characteristic down to his offspring.   6. **F**, A B2 star has a very short main sequence lifetime.   7. **F**, There are only a small number of stable orbits for planets orbiting binary stars.   8. **F**, An elliptical orbit would cause the temperature of a planet to fluctuate between extremes.   9. **F**, Europa is far outside the habitable zone. If Europa has subsurface liquid water, it is due to tidal heating by Jupiter.   10. **T**   11. **T**